The Sizing of Paper
Second Edition

A Project of the
Papermaking Additives Committee
of the
Paper and Board Manufacture Division

Edited by
Walter F. Reynolds

TAPPI PRESS

The information and data contained in this document were prepared by a technical committee of the Association. The committee and the Association assume no liability or responsibility in connection with the use of such information or data, including but not limited to any liability or responsibility under patent, copyright, or trade secret laws. The user is responsible for determining that this document is the most recent edition published.

Within the context of this work the authors refer to a number of specific manufacturers of equipment. This does not imply that these manufacturers are the only or best sources of the equipment. They are used by the authors as examples of the types of equipment available.

The presentation of such material is not intended by TAPPI and should not be construed as an endorsement of or suggestion for any agreed upon course of conduct or concerted action of any kind.

Preface

Since the first TAPPI Monograph on the "Sizing of Paper" was compiled and edited by John Swanson, the technology of sizing paper has significantly advanced. This advance has been especially dramatic in the area of cellulose reactive synthetic materials. With these products it is now feasible to produce sized grades of paper under neutral or alkaline pH conditions. Sized calcium carbonate filled paper is a good example of the results of developments in this area.

The technology of using rosin size and alum has also made great strides, e.g., dispersed rosin emulsions. Furthermore, application techniques, e.g., cataphoretic mobility techniques, now available to the industry have helped to maximize the effectiveness of all wet end additives. In addition powerful retention polymers are now available.

In 1981 the Papermaking Additives Committee decided that an up-to-date treatise on paper sizing was overdue. The project was broken down along product lines. The authors contributing to the project were chosen on the basis of their contributions and knowledge of a specific subject. Without exception the authors expressed a willingness to make the commitment needed to accomplish the task. Your editor was gratified with the response. Each author devoted many hours of his spare time to his chapter, and deserves our thanks for a job well done.

Certain guidelines were suggested. First and foremost, it was recommended that chapters be written to particularly benefit paper mill technical people, whose responsibility it is to maximize the performance of the sizing system. Not only should readers be instructed in the correct procedures, but they should also be informed on what to do when trouble occurs.

The chapter on "Mechanisms of Paper Wetting" is a monumental treatment of the subject. It deals in great depth with the subject and will be of great interest to technical people engaged in research. It is probably beyond the scope or interest of technical personnel responsible for day-to-day operations, however, it represents a profound contribution to a basic understanding of the subject and is therefore included in the Appendix section.

W. F. Reynolds
Editor

Contributors

R.M. Chad, 3M Co.
C.A. Schwartz, 3M Co.
Edward Strazdins, Consultant
R.E. Cates, Hercules Inc.
D.H. Dumas, Hercules Inc.
D. Bruce Evans, Hercules Inc.
Dale R. Dill, Protein Technologies International, Inc.
Kenneth A. Pollart, James River Corp.
T. Arnson, International Paper Co.
B. Crouse, Eastman Kodak Co.
W. Griggs, Eastman Kodak Co. (retired)
R.W. Kumler, deceased
J.M. Gess, Weyerhaeuser Paper Co.
John Swanson, Institute of Paper Chemistry (retired)
R. Wasser, American Cyanamid Co.
C. Farley, American Cyanamid Co.

Editor
W.F. Reynolds, American Cyanamid Co. (retired)

Contents

List of Figures

List of Tables

Appendix

1

Chemistry and Application of Rosin Size

E. Strazdins

Introduction

The purpose of sizing is to inhibit penetration of liquids into the internal structure of paper. If the liquid is polar, such as water, writing ink or lactic acid, then the sizing is accomplished by converting the hydrophilic fiber surface to a non-polar, hydrophobic surface by covering it in a controlled manner with a water repellent material. In the case of rosin sizing, the principal hydrophobic agent is aluminum rosinate which forms when a soap size (sodium rosinate) or free rosin acid is reacted with a hydroxy-aluminum salt at some point in the papermaking process. The soap size is water soluble and chemically very reactive. Formation of aluminum resinate occurs when the size is brought together with alum in paper stock at the wet end of the paper machine. Another type of rosin size is dispersion size, which consists of colloidally stabilized free rosin; the formation of aluminum resinate occurs when the paper web passes through the dryers.

The efficacy of the two forms of rosin sizes and the physical properties of the finished paper are related to the inherent differences in reactivities and rates of formation of the aluminum resinate.

Soap size is fully solubilized and highly anionic and is, therefore, extremely reactive with polycations. Any degree of interaction with an electrolyte alters its colloidal state, which has a direct bearing on sizing performance. The same considerations do not apply to dispersion size, which does not chemically react with alum during the wet-end addition. Because of its different physical state and different reactivity, dispersion size requires special handling and different application procedures from soap sizes. For these reasons dispersion size is regarded as a distinctly different sizing material, and the mill application is kept separate from that of the soap size.

Soap size is the oldest form of rosin size, and in its various forms it still represents the largest volume of rosin-based size used by the paper industry today. During introduction into paper stock, soap size may undergo major transitions in its chemical and colloidal state. A concise explanation of the

"

consequences of such changes requires a comprehensive overview of all mechanisms involved in sizing development. Therefore, this chapter will include a state-of-the-art summary of the theory of sizing for all forms of rosin size. A substantial amount of discussion will be devoted to the properties of soap sizes and to problems related to mill application. This kind of background is all-important to a papermaker for making judicious technical decisions in optimizing mill operation.

Sizing with rosin and alum has a long history, and an enormous amount of work has been done to cover its theoretical and technological aspects. Unfortunately, not all of the published work is useful for providing answers to the many questions which still surround the process of rosin sizing. Many investigations are limited in scope, or the experimental techniques used are inadequate for producing meaningful results. On the other hand, some aspects of sizing have been studied numerous times with basically the same conclusions reached each time. And yet some of the most critical phenomena are not well understood, and the fact is that sizing is often practiced without clear understanding of how many process variables ought to be controlled. The purpose of this chapter is to narrow this gap by consolidating all the well-documented information into a coherent theory that can help both the papermaker and the scientist tune their sizing system for best results.

Types of rosin sizes

For the manufacture of rosin sizes, three basic types of rosin are used. There are gum, wood, and tall oil rosins. Although they differ somewhat in chemical composition and physical properties, with some modification, all can be interchangeably used for the manufacture of rosin sizes. The proper choice of chemical modification is dictated by the type of rosin used and by the type and specifications of the finished size. These aspects are the responsibility of the size manufacturer, and the information is by and large proprietary. It is fair to say that no two commercial products are made the same way or possess identical physical and chemical properties. From the papermaker's point of view, it is important to discuss that information which has direct bearing on product handling, storage characteristics and application in the mill. The technological aspects of rosin size will be briefly discussed from that point of view.

Paste size is the most common and the most widely used form of rosin size. It is most stable, easy to transport, and easy to handle at the mill. Since this is the oldest form of size, most mills have the know-how and the equipment necessary for handling it. Although in the early days paste size was almost fully neutralized, now it is produced at 70% to 80% solids, which necessitates leaving 10% to 20% of the rosin acids in unneutralized form to optimize the stability and viscosity characteristics of the size. At high solids, paste size is very viscous at room temperature but assumes pumpable viscosity and possesses good water dispersibility when heated to 70° to 80°C. This type of size requires two-stage dilution: first with hot water to about 10% solids, followed by cold water to 3% to 5% solids.

Low viscosity sizes are similar to paste sizes in their composition. The difference is that other types of base, such as potassium hydroxide and viscosity depressants are used to render them fluid at room temperature at 50% to 60% solids (1). These sizes are suitable for low volume consumers, and they require no special equipment for emulsification and addition to paper stock.

Dry size is an easily soluble powder of fully neutralized rosin, which may be added to paper stock directly at any convenient point without predilution.

Dispersion sizes contain almost 100% free rosin which is colloidally stabilized and supplied to paper mills at about 40% solids. The properties and handling of this class of product are so different that they are covered in a separate section.

Specifications of the raw material rosin for these grades vary greatly as do the physical properties and the reactivities of these sizes in the papermaking process.

Influence of rosin on size quality

As a natural or semi-refined industrial product, rosin consists of a complex mixture of organic acids (2). The chemical and physical properties of rosin vary depending upon the amount and ratio of these acids. Some properties of the rosin, such as crystallization, are carried over to the size. For best storage and performance characteristics, most rosins have been subjected to some kind of chemical treatment which alters the ratio and nature of these rosin acids. Any form of rosin size is a complex colloid whose properties critically depend upon minor changes in the composition of rosin acids, the amount of neutral materials or special additives, size solids and added electrolytes. The performance of a size in the end use application likewise is affected by numerous fixed and transient variables of each mill system. Therefore, it is impossible to make categorical statements in regard to the efficiency of any type of rosin or any similar grade of rosin size except in a few instances. Experience has shown that high grade rosins produce sizes of good efficiency, while a low grade, such as B-wood rosin which contains oxidized materials and coloring matter, produces less efficient size. It is also known that introduction of additional carboxyl groups in rosin by fortification (3) using maleic or fumaric acids improves the efficiency of all sizes (4,5).

In converting rosin to any form of size, the manufacturer is always faced with crystallization of rosin and crystallization of the finished size as serious problems. These problems appear when rosin contains a high percentage of any one of the rosin acid isomers. The crystallization is overcome by suitable chemical modifications whereby the ratio of the rosin acids is balanced in such a way that propagation of crystal growth is inhibited. Other factors that favor formation of crystals in the size are high free acid content, high solids, contamination of size with seeds of insoluble crystals, and holding the size at a critical temperature, which for the high solids paste sizes is around 60°C.

Optimization of viscosity and the solids and free acid contents are important to economics and handling, but crystallization is the least forgiving factor since the problems that are created may be almost insurmountable to the papermaker. If a size is prone to crystallization and its crystal growth is unimpeded, the whole tank car or the entire contents of a storage tank may solidify. The most common crystal form is the 3:1, which consists of three moles abietic acid to one mole Na-abietate. This is a needle-like crystal, but its exact appearance is dependent upon the environment where the crystals have grown. The crystals are insoluble in water or alkali, and they cannot be destroyed or removed by any convenient means. Once the size has crystallized, it cannot be emulsified or pumped and must be removed from the tank by digging it out with pick and shovel. Another crystal type is the 1:1, which is very similar to the 3:1 type in appearance except that this form is slightly peptizable in hot water and will slowly dissolve in hot alkali. The 1:1 type crystals commonly occur in fortified sizes and can be eliminated by diluting the size with hot water, adding some alkali, and simultaneously agitating the size using live steam. Third, but much less common, are needle-like water soluble crystals which are the tri-sodium salt of the maleo-pimaric acid. These crystals may develop only in highly fortified sizes which are almost fully neutralized and where the Na+ - ion concentration is high. If size contains unreacted $NaHCO_3$, it might form platelet-like crystals which are water soluble. It goes without saying that water soluble crystals should be of no concern to the papermaker.

Crystals that form in dilute rosin size emulsions always are the 3:1 type. These crystals may grow quite rapidly and cause problems by plugging strainers in the emulsion lines and forming spots in the paper.

The inherent crystallization tendency of a size must be determined with certainty. This can be done by uniformly seeding the paste size samples with active seeds which have been generated by grinding insoluble crystals to a fine state and suspending them in water. A precisely measured quantity of the seed crystals is added to the sample and uniformly mixed in. The sample vial is tightly sealed and placed in an oven at 60°C. The samples are checked once a day for crystal growth by examining a thin film of smear under a low power microscope using crossed polarizers.

If the size is prone to crystallization, the seeded samples will develop traces of crystals within a day. Without seeding, it may take several weeks to develop signs of crystallization. In a tank car the crystal growth can be triggered by seeds of size from a previous shipment or by dehydrated size left on the coils.

Paste size offers the most challenge to the rosin chemist because of its extremely complex colloidal nature. Therefore, a complete optimization of all chemical and physical properties of a size can never be achieved, and one must always deal with a trade-off situation.

Properties of paste size

The effectiveness of a paste size as a sizing agent is mostly dependent on properties of the rosin and rosin additives from which the size is made.

Specifications of the size are set to primarily control the handling and emulsification properties of the size.

Paste size has a high solids content, usually 70% to 80%. At room temperature it is an extremely viscous paste, especially in the higher solids range. For adequate stability to crystallization and low handling costs, the following properties must be balanced: (1) free acid content, (2) total solids and (3) viscosity.

To minimize crystallization, the free acid content of the size must be kept low, although neutral sizes undergo enormous increases in viscosity due to the promotion of gel structure. The gel represents an orderly alignment of the sodium-rosinate molecules and is a quasi- crystalline structure. It is difficult to pump and has a low heat transfer coefficient. Heat transfer properties are not only important to the design of equipment for handling paste size but also to the amount of steam and time required to heat the size to unloading temperature. If the heat transfer is poor, overheating may cause dehydration of size on the coils, and that may completely arrest the heat exchange. All sizes, depending upon the solids content, show a viscosity minimum somewhere between 10% and 18% free acid, which manifests a transition in the colloidal state of the soap.

The situation is further complicated by stratification, which results in separation of size into two distinct layers. The bottom layer always possesses a more pronounced gel structure which has a lower free acid content, has lighter color, and is far more viscous than the top layer. Segregation of size in a mill storage tank obviously could create considerable operational problems and cause sizing fluctuations.

All these parameters are closely interrelated, and proper balancing of the colloidal and physical properties of paste size offers a continuous challenge to rosin size manufacturers.

Problems related to oxidation of rosin

The unsaturation of the three-ring hydrocarbon portion of abietic acid is expected to undergo most of the typical reactions of unsaturated hydrocarbons. Addition of oxygen to the diene system results in the formation of peroxides which decompose to form acid, alcohol, and aldehyde groups (7). If sized paper is exposed to ultraviolet light which accelerates these reactions, sizing may be completely lost. A build-up of peroxides in dry size may result in a spontaneous combustion since the heat resulting from oxidation cannot always be adequately dissipated when the dry size bags are stacked up in a warehouse. Manufacturers have experimented for years to stabilize rosin against oxidation. This is usually accomplished by disproportionating rosin or by adding antioxidants (8).

Foam in paper stock is another objectionable phenomenon which is promoted by many factors, some of which work in a synergistic manner. Oxidation of rosin or rosin size certainly is one foam-causing factor which must be taken into consideration. A size made from an oxidized poor quality rosin will be limited in performance.

Handling of paste size

The methods of handling different sizes are basically dictated by their physical and colloidal properties. Paste sizes are viscous at room temperature but thin out and become pumpable at temperatures above 60°C. These sizes are shipped in tank cars or special tank trucks. The railroad cars are equipped with heating coils, and the size is heated to a temperature where it becomes fluid enough for unloading. Viscosity problems can be aggravated by gellation and stratification. If the size has not been adequately stabilized against crystallization, a crystalline phase may form and settle out causing added unloading and emulsification problems. Size suppliers take utmost care to minimize these problems.

The best procedure for unloading a given size is usually provided by the supplier. As a general rule, low pressure steam should be used for preheating the tank car, since high pressure steam and its higher temperature can cause dessication of size on the heating coils reducing heat transfer.

Emulsification of paste size in the mill is performed using any of the several commercial emulsifiers that have been designed by the major rosin size manufacturers. To produce a stable dilute size, the emulsification must be done in two stages. In the first stage, the paste size, heated to 71°C to 82°C, is mixed with hot water to produce 10% to 16% solids primary emulsion. In the second stage, the hot primary emulsion is diluted under good agitation with cold water to 2% to 5% working concentration.

The same basic principle is used to dilute a paste size in the laboratory. Hot water is added stepwise with good mixing to achieve a perfectly uniform thin paste. When the size solids have decreased to about 10%, the rest of the dilution is done with cold water. Initially the heat is needed to disrupt the micellar or gelled structure of the size and to bring the size to a more or less completely solubilized state which is indicated by a sparkling clear appearance. Further dilution with cold water "freezes" this colloidal state preventing the unsaponified rosin from recombining to form semi-colloidal particles which may easily segregate out.

The appearance of emulsified size at 3% solids is dependent upon the free acid content and the amount of neutrals in the size. A properly emulsified rosin size normally is almost transparent, and in this state it is stable for prolonged periods. If the dilute size is cloudy, it may become further destabilized by hydrolysis, CO_2 absorption from air and water hardness, which results in the formation of insoluble Ca-soap. An unstable emulsion is strongly opaque and may cause deposits in storage tanks and coat out on the dilute size lines. In areas of hardness above 100 ppm, softened water must be used for preparing emulsion. The tolerance for heavier metal ions is much lower, and softening of water becomes even more important. The formation of Casoaps may be retarded by adding sodium hydroxide to the emulsion, which raises the pH and converts some of the unstable free rosin to soap form which acts as a stabilizer.

Standard analytical methods

Standard methods used for analyzing rosin sizes are listed in Table 1.1. For fortified sizes certain analytical tests require some modification. One such modification is in the determination of free alkali (TAPPI T 628), which is based on $NaHCO_3$ insolubility in alcohol. Since the tri-sodium salt of the maleopimaric acid is also insoluble in alcohol, the bicarbonate must be determined using a different method (9). Precipitation of the tri- sodium salt of the adduct during acid number of free acid titration should be of no concern.

The analysis for unsaponifiables also requires slight modification as reported in the literature (10).

Table 1.1 TAPPI Method for analyzing rosin size.

T 628 - Paste

Total solids	Ash
Free rosin	Alkalinity of ash
Combined rosin	
Unsaponifiable matter	...

T 629 - Dry

Total solids	Ash
Free rosin	Alkalinity of ash
Combined rosin	Free alkali
Unsaponifiable matter	Specific gravity

Fundamentals of rosin sizing

Development of sizing with rosin and alum involves many chemical and physical phenomena, all of which require due attention in order to keep the process properly controlled.

The necessary surface chemical conditions for development of hard sizing have been spelled out in three basic requirements (11,12):

1) Preparation of a size precipitate that has a potentially low free-surface energy and thus high water repellency,
2) Retention of size precipitate as uniformly as possible on the fiber surfaces,

3) Conversion of the sizing precipitate on the fiber surface to a stable, low free-energy film (Al-resinate) which stays fixed even though fluids are brought into contact with it.

Formation of the proper type of Al-resinate is the most critical phase in sizing. Soap size is water soluble and reacts as soon as it comes into contact with alum. Factors that influence the chemical and physical nature of the alumina floc also affect the size precipitate. Cellulose in water behaves like a polyanion. It bears a negative charge and has a high tendency to bind cations. The size precipitate that forms on reacting rosin with alum bears a cationic charge in the pH range of 4 to 6 and possesses high affinity for the electronegative cellulose fibers. This high affinity for cellulose is due to electrostatic attraction and also to the high escaping tendency of the semi-hydrophobic aluminum resinate from the aqueous phase (13). The precipitates formed from soap sizes consist of a mixture of aluminum mono-resinate and di-resinate and some free rosin (14,15). The amount of free rosin in the precipitate varies with the conditions at which the precipitation is carried out. Only a small amount of free acid is found in the precipitate if high amounts of alkali are present. Soap size precipitates are large clusters, and their distribution on fibers is far from being ideal for best sizing. The size distribution suffers even more if the stock contains large amounts of Ca^{++} or Mg^{++} ions or recycled alum. These electrolytes prematurely precipitate the size, and this highly agglomerated size is largely lost for sizing.

The strong attachment of size precipitates to fibers and fines provides a mechanism for retention. The high affinity of size precipitates for fibers also has some disadvantages. Some of the absorbed size is not capable of relocating during consolidation of the paper web and thus may strongly interfere with fiber bonding. Another problem is the high melting or sintering point of the precipitate. The aluminum mono- and di-resinates, depending upon the amount of moisture present, may lightly sinter (soften and reorganize) from about 120°C to 250°C or, if the proper moisture/temperature relationships are not met, may not sinter at all during paper drying (12,16). Inadequate sintering of the size precipitate precludes development of maximum sizing. Dispersion sizes cannot react with alum when added to fiber suspension because the dispersed rosin remains solid at the operating temperatures of the wet end. However, the soap or an anionic stabilizer that has been absorbed on the particle surface can react with hydroxy-aluminum ions to provide some retention mechanism. Effecting a good retention of dispersion sizes is a key problem, since alum alone cannot act in the most effective manner. It is usual practice to add retention aids.

Simultaneous action of heat and moisture are important elements in developing sizing with either form of rosin size. In the case of soap sizes, about 40% is the optimum moisture content at which maximum heat should be applied to induce some reorganization, or sintering, of the randomly formed size-alum precipitate (16,17). The combination of moisture and heat promotes reorientation of the hydrophobic rosin groups toward the air interface, which lends the size a more hydrophobic character. A properly scaled drying regime is very important for achieving maximum sizing with soap size. If too much heat is applied when the moisture content of the web is still high, the escaping steam may generate turbulence and debond the size precipitate from fibers. The debonded size agglomerates causes loss in sizing (18).

Development of sizing with dispersion sizes is influenced by the drying profile across the machine in a less critical manner, but it is clear that both heat and moisture work in a synergistic manner here also (19). Another inherent advantage of dispersion size is that it does not interfere with fiber bonding. The molten rosin migrates out of fiber cross-overs and develops sizing (aluminum resinate) when the sheet is almost dry. The delay in reaction with alumina also enables formation of a more uniform and a better-oriented size film, which translates into better sizing. Still another advantage is that dispersion size, which is almost 100% free rosin acid, can react and develop sizing with neutral hydroxy-alumina ($Al(OH)_3$). This means that the operating pH at the wet end is of less consequence, and sizing can be realized even in the neutral pH range. A limiting factor is the presence of a latent source of alkalinity, such as $CaCO_3$ filler, which supplies enough hydroxyl ions during paper drying to shift the paper pH to a range where rosin soap begins to form.

The principle of improving sizing by enhancing sintering of size precipitate can be restated in a different context. If the drying capacity of a paper machine is limited and the optimum temperature/moisture relationship cannot be capitalized on because the web temperature always remains on the low side, it might be worth considering a size that has been made from a rosin of reduced melting point (internally plasticized). Precipitates of such rosin will sinter at lower temperatures which will enable the necessary reorganization of the size during drying. It is theoretically impossible to optimize rosin size composition for all conditions. The temperature/moisture profiles of different paper grades and different machines vary greatly, and it is the machine operator's responsibility to adjust the temperature profile across the dryer cans to what appears optimum for a particular size.

The general working hypothesis always has been that a smaller particle size of the precipitated size will deliver more sizing (16). This requirement has been met by fortifying rosin with maleic acid (anhydride) or fumaric acid. The reaction products are tricarboxylic acids, which in the literature are referred to as maleopimaric or fumaropimaric acids. The added carboxyl groups are stronger acids than the abietic carboxyl (10), and this lends colloidally dispersed rosin a more electronegative character which translates into a smaller particle size (10). The smaller particle size of rosin is carried over to the size precipitates. Finer particle size precipitate provides more uniform surface coverage and better sizing. Since fortification also benefits the dispersion sizes, where the size spreads and reacts with alumina first when the paper web undergoes drying, it follows that other mechanisms, such as better bonding with alumina, may also be involved (18).

Chemistry of alum

The chemistry of sizing development with rosin is closely linked to the chemistry of alum. Factors that have an effect on alum also influence sizing (18). In principle some sizing with rosin can be realized using other metal salts that form insoluble hydroxydes (20), but alum is the best since it offers the right chemistry for developing good bonding of rosin to cellulose and produces the best sizing by critical test fluids. Another important consideration is the relatively low cost of alum. A detailed theory of sizing development with alum is given elsewhere (18,20). This treatise is limited to reviewing those aspects of alum chemistry that have a direct bearing on devising a useful working theory for the papermaker.

The chemistry of alum is extremely complex, and an enormous amount of work has been done to investigate its various aspects. As a result of the variety of techniques used, views differ as to which are the most likely aluminum ion species that do the work in colloidal interactions and which are the most logical reaction sequences in the wet-end application of alum. An analytical chemist always strives to work with a pure system that is well defined. A paper chemist is less fortunate since paper furnishes are of an extremely complex and variable nature which cannot be defined in terms of precise analytical parameters. The approach here must be semi-empirical, and techniques must be used that produce meaningful results from the practical point of view.

The trivalent Al+++ ion for a long time has been thought of as the principal species which does the job in interactions with anionic polyelectrolytes, such as rosin or cellulose. More recently many researchers have concluded that the more complex hydroxy-aluminum ions are the working species and that the action of alum in papermaking is largely related to the formation of polymeric alumina as an end product which bridges particles and neutralizes the electronegative charge of stock components (20, 21). A more clean cut delineation of the significance of these various mechanisms has been hampered by the lack of a technique that would measure the actually available charge of the various alum compositions. Control of alum chemistry in papermaking has been administered by establishing strictly empirical correlations of some parameter with the on-machine performance. The two most widely monitored parameters are *total acidity* and *degree of neutralization* of alum. These two parameters are interrelated to the extent that both in some way tie in with the most desirable form of alum for the particular condition.

By purely empirical experimentation, it has been established that anionic polyacrylamide dry strength resins perform best if alum, which is added for resin retention, is neutralized to about 30% to 50% by adding base to the stock (22). A recently devised titration technique, which enables a quantitative assessment of the available cationic charge of alum over a wide range of compositions, has shown that alum possesses maximum cationic charge at about 30% to 50% neutralization (19, 23). It was also shown that the available charge of alum depends upon many other factors, such as stock temperature, aluminum concentration, presence of polyanions, and even purely mechanical factors such as rate of mixing (23). This study has shown that the available charge of alum is a surprisingly low 1.8% of the formal charge of the trivalent Al+++ when the charge is titrated on a preformed alum floc. If the floc is subjected to shear or the base used for neutralization of Al+++ ion is added after alum has come into contact with the anionic poly-electrolyte, then the titratable charge increases to about 8%. With decreasing Al concentration, the maximum of cationic charge shifts to a lower formal degree of neutralization because at a low Al concentration, the pH is higher and there is a corresponding increase in the degree of hydrolysis. This says that the optimum pH for sizing shifts to a lower pH at higher alum dosages. This explains why the optimum pH is lower when sizing experiments are done by recycling white water (24).

Relatively short exposure of alum to heat brings about a significant reduction in charge, presumably as a result of Al-ion oxolation which causes formation of a more rigid network where the cationic sites become increasingly inaccessible for ion exchange (23). For example, 70% of cationic charge is lost if the hydroxy-aluminum salt is kept at 65°C for 10 minutes. Since a major proportion of alum is continuously recycled through the white water loop, it is no wonder that sizing and retention problems arise when stock temperature rises. The fact that the titratable charge of alum floc is only about 1.8% may not be totally surprising. The most rational explanation is that the sulfate by suppressing the electrokinetic charge of hydroxy-aluminum ions (20) brings about almost instantaneous formation of a dense floc, whereby a major proportion of the remaining cationic sites become "buried" and are unavailable for ion exchange. Strongly coordinating anions, such as citrate and hexametaphosphate, reduce the cationic charge of alum even more drastically at concentrations of 0.1 to 0.2 and 0.05 mols/Al, respectively, and reverse the charge to negative at higher concentrations. This explains very well the losses in sizing when citrate is present (25).

This information has several practical implications regarding the use of alum in the mill. If mixing at the point of addition is poor, alum will form floc before adequate distribution and interaction with stock components has taken place. Equally important is the avoidance of heat. If the wet-end conditions are such that the cationic activity of alum may be impaired, some of the most effective counter-measures are the use of more alum keeping contact times short and adding cationic polymers to offset the cationic deficiency.

The rationale for controlling the degree of neutralization of alum in dry strength resin application is quite simple because the anionic polyacrylamide requires the most cationic form of hydroxy-alumina to be retained on electro-negatively charged cellulose. Rosin sizing, by comparison, is more complex. Nevertheless, the same general trend also prevails there. Extensive mill studies (25a) have shown that maximum sizing and maximum retention is achieved if alum is 30% to 40% neutralized.

To take advantage of this chemistry in rosin sizing, it is necessary to exercise good process control. The stock alkalinity and the amount of alum must be carefully monitored. The control of total acidity is another way of accomplishing basically the same goal. The problem in either case is that the maximum cationic charge of alum is a function of Al concentration as well as degree of neutralization (23). Other important variables are retention, degree of recycling of machine tray water, pH at different points of stock flow, point of size addition, pH before alum addition and the pH history, mixing, contact times, water hardness, and the presence of other electrolytes. The individual and combined effects of these variables are too complex to enable oversimplified predictions. The bottom line is that the optimum degree of neutralization and the optimum total acidity of the furnish are parameters that must be independently established for each mill system.

Sodium aluminate is a non-acidic source of alumina that has been used for years in the rosin sizing process as a partial substitute for alum. The rationale for using sodium aluminate is to supply to the system a given amount of alumina without excessively increasing the sulfate concentration. Several benefits have been claimed to accrue from the use of aluminate. One is the ability to operate at a higher pH and others are less corrosion better paper strength, better sizing, and better retention (26). The application of sodium aluminate is based on empirical background without having a well-documented theory of how the chemistry works. A major reason for the lack of better understanding of the physical- chemical phenomena involved has been the unavailability of a suitable technique that would allow quantitative assessment of the available cationic charge of the alum-sodium aluminate interaction products. On the basis of recent theoretical work (23) which has provided some information on the available cationic charge of the various forms of alumina formed from aluminum sulfate and aluminum chloride, it might be possible to conjecture on the charge relationships of the aluminate-alum reaction products. The cationic charge of alumina that is formed in the absence of sulfate ion or at low sulfate/Al ratios (20) is much higher than that of alumina formed from alum. It seems logical that the mixtures of aluminate and alum will have greatly reduced sulfate concentrations; therefore, the available cationic charge should be accordingly higher than that of alumina formed from 100% alum. This factor should be particularly significant at higher pH's; i.e., at higher degrees of neutralization of Al+++ ion.

Alum demand always has been a debatable topic. In papermaking, alum serves multiple purposes. It develops sizing with rosin, it provides mechanisms for retention and charge neutralization, and it provides active bonding sites for common retention aids that have anionic pendant groups. All these mechanisms have some inherent alum demand. As discussed earlier, there are many factors which adversely affect the cationic charge of alumina and thus greatly increase the demand for alum.

Soap size requires more alum than dispersion size. As a general rule, a typical alum to size weight ratio for the two size systems is 1.5:1 and 1:1, respectively. The minimum amount of alum recommended for sizing well accords with the theoretical amount of aluminum sulfate needed for the formation of aluminum mono-rosinate, which has the highest contact angle with water (27). In mill practice generally more alum is used for a number of reasons, some of which were indicated earlier. Many mills use alum for pH control. If the incoming stock is alkaline or the process water has a high bicarbonate content, very high levels of alum may be required to bring the system to proper pH. Another reason for excessive alum usage is that agitation in many mills is grossly inadequate for rapid distribution of alum in paper stock before the formation of alumina floc has commenced (23, 28). If alum is allowed to form a precipitate before distribution to potential adsorption sites on furnish components, much of the cationic action and the capacity of physical bridging is lost. As pointed out earlier, hot stock is another important factor in reducing the charge of alum. Other factors that influence alum demand are type of size, polyelectrolytes. Therefore, the amount and the best method of using alum must be independently established for each mill system.

Experience has shown that the alum demand can be significantly reduced by using a mixture of alum and sulfuric acid. It is to the mill's advantage to avoid overloading the system with alumina. Sulfuric acid is not only more cost effective for pH control but also acts as a pH buffer in delaying the premature formation of alumina precipitate in paper stock (28).

Effects of electrolytes

The effects of electrolytes in rosin sizing becomes a hotly debatable subject if one tries to critically review all the published information. Sizing is a complex process where any deviations in some critical variable may produce totally different results. Since the published information on this subject is not consistent, an attempt will be made to overview the main effects of electrolytes in the context of reasonably documented theories.

In regard to anions, the early theory of Thomas (29), who ranked anions in the order of increasing coordination tendency with aluminum, still is useful since the suggested coordination tendency correlates with the adverse effects of anions on alumina charge (23).

The reduction of alumina charge does increase in the order given by Thomas, i.e., nitrate (chloride) sulfate citrate. To this must be added the polyphosphates (hexameta-phosphate) which are the worst offenders (23). While Thomas' theory was correct to predict the effects of anions on cationic charge of alumina, the complete mechanisms of the action of coordinating anions had never been fully understood since there are other factors which are equally important. The point is that the coordinating anions, such as citrate, influence the sizing and alum chemistry by more than one mechanism (12). For example, the citrate, if present in excessive amounts, destroys sizing (25) of regular pulps. The situation is opposite if highly purified (alpha-cellulose) pulps are used. There a small amount of citrate (0.1 mole per Al) produces a major improvement in sizing (12). The explanation is that a small amount of citrate renders the alumina more heterogeneous and reduces the sintering point of the size precipitate (12) which is an important mechanism in developing sizing. Consequently, for pure pulps a small amount of a coordinating anion is to be considered a desirable "impurity." For normal commercial pulps, which already contain various kinds and various amounts of contaminants such as saccharinic acid from the peeling reaction during alkaline pulping, the presence of additional coordinating anions can only be detrimental to sizing (12). Similar arguments apply to the sulfate ion, although the mechanisms involved are different. A small amount of sulfate ion is desirable inasmuch as it moderates the strongly cationic charge of alumina (20) (formed from $AlCl_3$), but too much sulfate is detrimental because it excessively suppresses the cationic charge of alumina and induces rapid floc formation before the aluminum ions have had a chance to distribute evenly to the adsorption sites (23).

An excessive build up in ionic strength suppresses the electric double layer and electrostatic (coulombic) force interactions. The nuclei of positively charged size precipitates cannot adsorb on fibers rapidly enough to compete with the rate of growth to large particles, which are not a desirable physical state of the size for developing maximum sizing (16, 23). In summary, excessively high levels of electrolytes are detrimental to sizing, although the mechanisms of action and critical concentrations of each ionic species are different.

Cations influence sizing by different mechanisms and these also are not fully understood. Ca^{++} ion is the most common cation, besides aluminum ions, that deserves consideration. In large concentrations (100 to 300 ppm as $CaCO_3$), calcium can be quite detrimental to sizing primarily by the mechanism of prematurely aggregating the dilute rosin size. The calcium soap that forms is an ineffective size (16, 18, 30). On the other hand, it has been found that some calcium is beneficial to sizing (31). It is a well-known fact that sizing problems appear in alpine valleys where the water is derived from melting snow and the process water is very soft. It has been recommended that calcium be present in an amount equivalent to one CaO to one mole of Al_2O_3 (11,31). Other investigators have found that Ca^{++} ion and other divalent cations increase the electrokinetic charge (mobility) of alumina (32), which is consistent with the theory that divalent cations benefit sizing. Under situations where the mill water is too soft, it is common practice to add hydrated lime to the stock before rosin size and alum are introduced (11). The amount of lime used generally amounts to a few pounds per ton of paper.

The effects of electrolytes have received little attention from research works, and one can only speculate on the more likely mechanisms by which the alkaline earth metals work. In doing this, one should recognize the fact that the incorporation of some electrolyte usually also involves buffering the furnish around a slightly acid pH (5-7). This by itself is an important factor for converting the fully solubilized rosin soap to a colloidal state where some free rosin is liberated. Since divalent Ca^{++} ion is instrumental in reducing the electronegative charge of the rosin size micelles, the formation of colloidal rosin is further accelerated. Alum precipitates formed from this form of rosin size sinter at lower temperatures, which is a desirable feature in sizing development.

The preceding discussion has suggested the basic modes of action of some cations. Other cations are expected to act in a manner as suggested by the Schulze Hardy rule. Provision of the basic modes of action of a few electrolytes provides some working theory to the papermaker on how to tackle his sizing problems and how to change the mill-operating procedures for maximum benefit.

Summary of surface-chemical principles of sizing

The principal surface-chemical phenomena of sizing with a soap size and alum are summarized by results of a laboratory experiment shown in Figure 1.1. The amount of sizing and alumina retention (Al_2O_3) are plotted against the pH at which the handsheets were prepared.

At low pH (3.5) the rosin retention is adequately high, but not enough alumina is present to form aluminum rosinate. At pH 4.5 to 5.0 both rosin and alumina are present in the right amounts and hard sizing develops. In this pH range the trivalent Al^{+++} ion is converted to hydroxy-aluminum ions which have a strong cationic charge. The size is precipitated as aluminum rosinate, and some free rosin acid is liberated by hydrolysis of the soap. The precipitate is strongly cationic and rapidly adsorbs on fibers and other finely dispersed stock components. When the paper web enters the dryer section and reaches a high enough temperature, the precipitate sinters, and the free rosin completes its reaction with hydroxy- alumina to form aluminum rosinate.

In laboratory maximum sizing is obtained in the pH range of 4.5 to 5.0. Under mill conditions where the stock is hot and alum concentrations are higher because of white water recycle, the optimum pH for sizing declines to about 4.1 to 4.3.

When pH exceeds 5.0, sizing rapidly declines. In the pH range above 5.0, some residual soap remains which contaminates the sized surfaces converting a low energy hydrophobic surface to a high energy hydrophilic surface which is readily wet by polar liquids and supports capillary penetration of the test fluids. The amount of soap present can be crudely estimated from the log (Ab–/ABH) curve (16).

The situation is different with dispersion sizes. Here the latent acidity of the free rosin acts as a buffer and the aluminum rosinate forms without any contaminating soap being present when the paper web reaches adequately high temperatures in the dryer. For this reason, hard sizing with dispersion sizes can be obtained over a broader pH range provided that other critical conditions are met.

This type of chemistry offers the papermaker several options for achieving satisfactory sizing with rosin in furnishes that even contain moderate amounts of $CaCO_3$ filler (33). Further discussion of this aspect will be made in the section on mill problems.

Mill factors

In the preceding sections the emphasis was placed on discussing the most important aspects of rosin sizing from the theoretical point of view. The idea was to provide the papermaker with some basic understanding of the complex nature of sizing in order to help him find solutions to problems that must be dealt with on a day-to-day basis in his own mill. Discussion of mill factors will entail a more specific application of the theory of sizing development towards explaining the miscellaneous problems that have been widely commented on in the sizing literature.

If one seeks a reasonably clean-cut answer to a problem by surveying the existing literature, the prospects are rather dim. Most of the trade literature represents specific case histories and isolated laboratory experiments, the critical parameters of which are never adequately defined. As a result, the information is not pertinent to particular systems and is therefore of limited value. It seems more productive to relate the causes and solutions of mill problems to some well understood physical-chemical principles. This approach does require some arbitrary choices in recognizing the pertinent literature and stressing the merit of certain analytical techniques, but it provides more understanding of the chemistry of sizing development, an objective that has eluded papermakers for generations.

Pulps

Fiber furnishes vary widely in the ease of sizing with rosin. In general, the purest pulps, such as rag or high alpha-cellulose pulps, are most difficult to size, while unbleached pulps can be sized much more easily (34). Unbleached kraft can be sized more easily than unbleached sulfite. While the ease of sizing is not related to rosin retention, the evidence is overwhelming that the differences in sizability are related to certain chemical and physical factors. Pectins and chemicelluloses have been found to directly contribute to the ease of sizing (35). With regard to kraft pulp, the wood species, cooking and bleaching conditions are recognized as important variables (37). Bleaching, and especially bleaching combined with alkali extraction, reduces sizability. Low brightness bleached pulp is easier to size than is a fully bleached pulp. Other studies have shown that bleaching of kraft and sulfite with chlorine or chlorine dioxide and acidification lowers the sizability of sulfite, while hypochlorite and alkali extraction improve the sizability of sulfite (38). Unlike sulfite, kraft pulp sizability decreases by alkali extraction and hypochlorite, chlorine or chlorine dioxide bleaching. Peroxide bleaching is said to be the least damaging, and washing the pulp after bleaching improves the sizability regardless of the bleaching treatment.

One could go on and on reciting all the work that has been done by numerous investigators, but the bottom line is that the conclusions are not definite enough to be applicable to a specific set of conditions. A non-critical review of literature is of very little value, while critical review is not feasible because of the lack of adequate identification of important parameters. The best one can do is to jump the gap and discuss these and related factors by making use of more advanced surface-chemical research in the area of rosin sizing.

Pulps vary from highly purified celluloses to unbleached pulps that have been poorly washed and are loaded with strongly anionic lignins and other anionic contaminants such as bleach residues or materials brought back into the system by white water recycle. Certain natural colloids that occur in unbleached pulps are beneficial to rosin sizing (12), since they reduce the sintering point of rosin-alum precipitate and provide for better size distribution and orientation during the paper drying step. Such "desirable" colloids are the saccharinic acids that result from alkaline pulping and anionic hemicelluloses that occur in unbleached pulps. In general these compounds are of the hydroxyacid type. These materials are indigenous to the unbleached and less purified pulps. The hydroxyacids complex with aluminum ions, which produces size precipitates of lower sintering points, a mechanism that promotes better sizing.

If the pulp is pure cellulose, such as alpha-cellulose, aluminum resinates form that do not adequately sinter (reorganize) and the size does not develop maximum hydrophobing properties. Hence, pure pulps are difficult to size with the soap type rosin size. However, good sizing can be achieved on adding a suitable amount of a hydroxyacid, such as citric (12,18).

From these considerations it should become obvious that in proceeding from poorly washed unbleached pulps to highly purified bleached pulps, a whole host of intermediate situations exists.

What has been said so far primarily applies to the soap type rosin size. Soap size forms aluminum resinate at the wet end before the paper web is formed. With dispersion sizes the situation is grossly different in that the rosin is retained in the sheet in unreacted form and the aluminum resinate forms when the paper web has reached the melting point of rosin which occurs in the dryer section. At this point the rosin spreads and reacts with retained hydroxy-aluminum to produce a well-formed hydrophobic aluminum resinate film. This mechanism does not require the intervention of "chelating colloids" (12). By this theory, the dispersion sizes appear the product of choice for bleached pulps that are difficult to size with the conventional soap type sizes. Mill experience seems to support this conclusion. In bleached pulps the dispersion sizes far outperform the soap types, while in unbleached pulps the performance differences are smaller.

Freshly cut wood is another factor that can adversely influence sizing. Sizing problems with freshly cut wood are aggravated in early spring when spring wood is forming. These problems may be related to the presence of unsaturated fatty acids which reduce sizing. Seasoning of wood for as little as 30 days may overcome this difficulty. Aging causes the fatty acid content to decrease.

An entirely separate problem is the difficulty of sizing groundwood. The causes of poor sizing of groundwood are not fully understood but may be related to the much larger surface area of the finely ground wood chips and the absence of desirable hemicellulosic material on the fiber surface.

Ion exchange capacity is another important factor that influences pulp sizability. Pulps with high carboxyl-group content are easier to size. Possible mechanisms are ion exchange with the soap size, which results in liberation of colloidal rosin, and ion exchange with alum which provides for better bonding and more uniform size distribution. For example, it has been shown that better sizing develops if the pulp has been washed with acid, i.e., the carboxyl groups have been converted to the free acid form.

Refining

The effects of mechanical action on the fibers in regard to sizing can be logically divided into several phases (11):

1. The question of sizing highly beaten fibers versus unbeaten fibers.
2. The effects of mechanical action on the absorbed size precipitate.
3. The effect of beating on the mechanical structure of the finished sheet and the influence of structural changes on sizing.

Sizing works primarily by rendering the exposed surfaces more hydrophobic. As the whole surface area of the furnish is enlarged, more size is required to achieve adequate protection. On the other hand, moderate beating improves the sizing results (39). The influence of beating on sheet structure and sizing has been studied at considerable length. The conclusion is that the effects of modifying paper structure are not simple (39,40). Moderate beating improves sizing by increasing the density of paper and decreasing the pore radii, thus retarding penetration. With further beating there is less change in density, porosity and pore size; however, the surface area increases and more size is required to achieve adequate surface coverage. It has been amply demonstrated that refining generates fines or loose fibrilla material that have large surface area and adsorb proportionately larger amounts of the sizing material (41,43). Beating also delaminates the fiber wall, making the fibers more conformable. This mechanical action produces intrafiber channels and partially shifts the penetration of a polar liquid from interfiber capillaries to intrafiber channels. These structural changes are bound to have varying effects on the sizing responses by various sizing tests.

Beating action also brings about some release of soluble or partially soluble residual cooking liquors, bleach residues and peptizable hemicellulosic materials, which not only increase the cationic demand (C.D.) of the furnish but also have major influence on the sizing response (18,45). Further details of the chemistry involved were discussed in the theoretical section on sizing development.

The results discussed so far are based on sizing the stock after refining. If the size is precipitated before refining, the effects of heavy shear on the structure and distribution of size precipitate must also be considered.

As a general rule, heavy shearing action is considered detrimental to sizing. This may be due to brushing off the size precipitate from fibers, which induces size aggregation and opens new surface areas that remain unprotected. The dispersion size is also shear sensitive and every effort must be made to reset it back on fibers before the web is formed. On the other hand, the dispersion size, because of its delayed reaction with alum (during paper drying), can adequately redistribute on fiber surfaces, but the soap size precipitates cannot because their sintering temperatures usually exceed the drying temperature of paper.

Hot Stock

Stock temperatures vary from mill to mill and range from room temperature to near 93°C. High stock temperature results either from deliberate injection of steam or inadvertantly from the operating conditions in the mill. The usual reason for injecting steam into the stock is to increase production through faster drainage and drying. Unintentional hot stock results from closed systems which retain heat, high temperatures in waste and broke repulping systems, and seasonal variations in the temperature of mill water.

It is well known that high temperatures are detrimental to rosin sizing and a large amount of work has been done to assess these effects. The most comprehensive treatise on hot stock problems has been given in Sizing Technical Highlights (44), but the most accurate explanation of these effects has come from theoretical work on alum chemistry (23), since factors that affect the chemistry of alum also affect sizing with rosin. Briefly, the active aluminum ion species that are responsible for retaining rosin and developing sizing are the 1/3 to 1/2 formally neutralized aluminum ions which associate to a hydrated, positively charged alumina polymer. The active or useful species for precipitating and retaining rosin size are the same that serve best to retain anionic polyacrylamide dry strength resins. When the stock is heated or allowed to stand, further crosslinking and polymerization of the hydroxyl groups set in, termed "oxolation", by eliminating water between neighboring hydroxyl groups. As this dehydration proceeds, the floc shrinks and aggregates, which results in poorly distributed size precipitates and greatly reduced cationic charge of the alumina (23). Heating at 65°C for 10 minutes reduces the available cationic charge of alumina by 70%. The losses in sizing may be even more dramatic (44).

Some suggestions for improving sizing at high temperatures are offered below:

1. Add the alum to the system as late as possible. Keep the contact time of the alum rosin precipitate as short as possible.
2. Increase the alum level somewhat to make up for partially dehydrated or aggregated floc. Do not rely on pH alone to adjust the alum level. (For a fixed amount of alum, hot stock develops a lower pH than cold stock).
3. Lower the machine pH, i.e., operate at a lower degree of alum neutralization, since this keeps the floc more charged and less subject to dehydration and aggregation.

4. Reduce the sulfate buildup by adding fresh water. (High sulfate ion concentrations accentuate the effect of hot stock because they partially neutralize the positive charge of the floc, thus permitting them to aggregate).
5. Use fortified rosin size which produces smaller particle size and is more resistant to temperature effects than non-fortified sizes.
6. Use dispersion size which entails different (delayed) mechanisms of forming the aluminum resinate and thus is less sensitive to the adversities of hot stock on alum chemistry.
7. Keep steam jets before, rather than after, precipitation of the size. The temperature near the steam port will be much higher than the average stock temperature.
8. Lower the stock temperature if at all possible. Even a little fresh water may help.
9. Utilize dry strength resins and retention aids to improve drainage rates and thus increase production. The dry strength resin permits reduced refining, but holds the strength level and sizing.
10. Add trim alum or retention aids to set unretained size precipitate, since on recycle the unretained precipitate will lose cationic charge and extensively aggregate.

Another factor that deserves some consideration at hot stock conditions is the pulp itself. It has been reported that elevated temperatures have more effect on the sizing of bleached sulfite than on either unbleached sulfite or unbleached kraft (25). These effects may be ascribed to the absence or presence of "chelating colloids" (18), which are expected to retard the shrinkage of alumina flocs (and consequently reduce the aggregation of size precipitate) at high temperatures. Sizing may also be aggrevated by seasonal adjustments in furnish, such as use of more groundwood to compensate for formation problems during hot weather (25), or by an increased load of dissolved organic matter (humic and fulvic acids) in the process water (46). Sizing problems are interrelated most of the time; by solving one problem, quite often a new problem is introduced. Therefore, the papermaker must be fully aware of these effects and choose the proper trade-off situation.

Process water

The presence of dark color in process water is a good indication of dissolved organic matter such as humic acids. The humic acids affect sizing adversely by reducing the cationic charge of alumina (46). This problem can be offset by using more alum or adding a small amount of cationic polymer as a charge assist. Since cationics in combination with alum retain these dark colored organic materials in the sheet, one may experience some loss in brightness. Similar considerations apply to other polyanionics, such as residual lignins.

The effects of inorganic ions, such as calcium, on sizing with soap size were discussed at some length earlier. In general any appreciable amount of divalent or trivalent cations is expected to be deleterious to sizing with a soap size by causing premature aggregation of the size before it is contacted by alum. However, a small amount of calcium is very beneficial to sizing in laboratory if deionized or distilled water is used for stock makeup. This might be related to inducing more hydrolysis of the soap, which results in the production of more free rosin and lower sintering points for the precipitated size.

Furnish ingredients

In principle, all furnish additives have some effect on sizing. While cationic polymers may have beneficial effects on sizing, most other additives have negative effects to varying degrees and by different mechanisms. Some additives may enlarge the surface area of the furnish, others may alter the surface-chemical nature of the surfaces or interfere with the chemistry of alum. If the amount is small the effect may not be significant, but situations may exist when sizing is lost completely. Adverse situations may be heightened by closure of the white water loop. It is up to the papermaker to foresee the development of these situations well in advance and be prepared to take corrective measures.

Fillers: All fillers reduce sizing to some degree. In the case of clay and titanium dioxide the negative effect may be attributed to enlargement of the total surface area and to introduction of certain anionic contaminants, such as polyphosphates. These polyanions have a profound adverse effect on the alum chemistry, as discussed in more detail earlier. Of some consequence is the fact that the wet-end additive adsorbtion is always proportional to the surface area of a furnish component. Since the fillers and fiber fines have large surface area a substantial proportion of the size adsorbs on fines and fillers (41, 42, 47). To maintain maximum sizing and to minimize degradation of the size qualities on recycle by aggregation, every effort must be made to maintain high first-pass retention (18). This becomes of particular importance if stock is hot.

Calcium carbonate has the most profound effect on rosin sizing, because of its ability to neutralize alum and maintain pH in the range where Ca-soap begins to form (7-8). The situation has changed with the advent of introduction of the dispersion sizes. As indicated in the last section on "Use of Rosin in Neutral Papermaking," there are some alternatives open to the papermaker for maintaining satisfactory sizing with rosin under these conditions. Use of wax sizes along with rosin offers an additional option for improving sizing in the presence of calcium carbonate filler (48).

Calcium silicate is alkaline and is expected to interfere with sizing to some degree if used in appreciable quantities. Other fillers such as gypsum and talc are neutral and will affect sizing only by enlarging the total surface area of the furnish.

Polymeric Additives: A broad spectrum of natural and synthetic polymers is used in papermaking for a number of purposes. Due to specific performance requirements, their chemical composition, charge, molecular weight and physical-chemical behavior are tailored to suit the specific needs. All additives, regardless of their intended function, will have some effect on rosin size performance. The mechanisms of action vary and frequently are not well understood.

Cationic polymers always act by the charge neutralization mechanism; if their molecular weight is high enough the mechanism of physical bridging of particles will also be a contributing factor. In that sense the two mechanisms always act in a collaborative manner in effecting retention of all furnish components. The fines and fillers adsorb a major proportion of the rosin size precipitate (41, 42). Since exposure to hot stock degrades the size qualities, it is of utmost importance that the rosin-rich fines are retained in the sheet on the first pass. This objective is achieved by proper combination of cationic resins (promoters) with high molecular weight retention aids (66, 67). The size distribution in the Z-direction tends to parallel fines, and their distribution is not uniform through the sheet. Usually the size is deficient on the wire side and rich on the felt side of a fourdrinier-made sheet (49). On testing the sizing, two-sidedness is often evident. The extent of two- sidedness is a function of machine speed. Other factors that affect two-sidedness are total fines content, fillers, retention aids, furnish charge and basis weight.

Cationic resins enhance sizing also by improving orientation of the size. This is deduced from the fact that in the presence of cationic resins it is possible to size paper with soap size at pH 6 to 7, in a range where neutral alumina forms and where by the normal practice, sizing would have been very poor due to formation of rosin soap. Thus cationic polymers serve as a partial substitute for alum in pH ranges where alum by itself cannot deliver enough cationic charge. Cationic wet strength resins also contribute to sizing by retarding swelling of fibers and fiber crossover areas.

Anionic acrylamide-acrylic acid copolymers which are used for dry strength purposes in conjunction with alum also produce significant improvements in sizing with rosin (50). The most likely mechanism is improved retention of fines and, perhaps, some enhancement of size distribution.

Numerous studies cover the effects of natural or modified natural polymers on rosin sizing. This includes starches, natural gums and carboxymetyl cellulose. Cationic derivatives always produce some increases in sizing, most likely by improving size retention. Their anionic counterparts or the unmodified neutral polymers are subject to mixed opinion. A detailed account of the published information would be of little value, since most of this work entails laboratory experiments without giving adequate details about important parameters that influence sizing. Literature references 51 through 55 cover the most interesting studies on the application of natural polymers in rosin sizing. One product that stands out with producing significant increases in sizing is a 20% to 40% converted oxidized starch (51). Oxidation introduces some carboxyl groups which renders the starch more reactive with alum, which in turn may help sizing by several mechanisms. A solubilized starch after crosslinking with alum becomes an effective retention aid for size and fines.

The reaction product may also reduce the sintering temperature of the size precipitate which provides for better size distribution. Also it may act as a stabilizing protective colloid for the size during precipitation with alum, resulting in smaller particle size.

Another class of natural polymers that improve sizing with rosin are the proteins. These polymers act as stabilizers or protective colloids in the preparation of free rosin dispersions of small particle size. Bewoid size (56) and Prosize (57) are examples of successful sizing agents. Addition of animal glue or casein to stock along with soap size, before alum is introduced, always produces beneficial effects (58-61). Numerous studies in this area have resulted in one general conclusion, i.e., that all proteins to varying degrees act as protective colloids for the size, providing finer particle size and better size distribution on fiber surfaces, which always translates into more sizing. These earlier theories on the role of proteins are perfectly consistent with the present views, which are supported by a more direct surface-chemical evidence (16, 18) regarding the significance of smaller particle size.

Other Furnish Ingredients: Other ingredients may be beneficial or harmful, depending upon their nature and the amounts present. Defoamers are surface active and can have adverse effects on sizing if present in excessive quantities. This happens if too much defoamer is added to combat difficult foam problems. It is far better to get at the cause of a foam problem than to try to cure the result. Some of the most common ways of reducing foam include: eliminating sources of air (leaky pumps, cascades), allowing air to leave the system, controlling total acidity, adding alum as soon as size is introduced, using steam to break foam, controlling bleach residues, and using more water in stock. Generally, the foaming tendency is maximum at pH 6 and again at pH 4. This is because rosin soap (at pH 6) and excessive amounts of free, unbound rosin (below pH 4) are good foam stabilizers. Use of defoamers may also help sizing if improved sheet formation results from deaeration of the stock.

The effects of dyes have received less attention. Most dyes have very little or no effect on sizing, unless they are used in very large quantities (62).

Physical factors

The quality of sizing is significantly influenced by a large number of physical factors. Basis weight, density, inter-fiber bonding, formation and fines distribution are some of the key factors that deserve consideration. The basis weight or thickness of paper has received a lot of attention, the conclusion being that the liquid penetration time through paper varies with the square or the cube of the thickness (62, 63). The precise relationship is a function of the type of test used. In regard to sheet density, the results are strongly interrelated with the type of size used and the level of size (16). Wet pressing significantly increases sizing values (16, 42), fortified rosin sizes responding in somewhat different manner than the non- fortified sizes.

Development of more interfiber bonding frequently coincides with conditions that are more favorable for development of sizing. A typical case is the behavior of high free rosin sizes, which spread and redistribute on fiber surfaces during paper drying, thus allowing more fiber bonds to establish. Loss in formation always affects sizing adversely, since the thin spots get penetrated much more rapidly than the dense areas. This shows up as pinholes during the penetration tests.

Drying conditions have long been recognized as a key factor in realizing maximum sizing. The aluminum resinate precipitate must sinter to some degree to achieve the proper orientation on fibers. This only occurs if the proper moisture/heat relationship is established (12,18). The web must not be too hot while the moisture content is high, because live steam shears off adsorbed size precipitate and induces extensive agglomeration of the size. Both factors have adverse effects on sizing. It is agreed that paper drying must be conducted in a programmed way: cool early driers, hot center driers (maximum heat when web moisture has dropped to about 40%) and cool late driers (64, 65). In this context one must bear in mind the all-important fact that for different forms of rosin size the optimum drying regime will be different. Soap size precipitates largely consist of aluminum resinate which never adequately sinters (16) and whatever reorganization does take place during paper drying is critically dependent upon having the right moisture/temperature relationship. Both factors work together to effect the necessary orientation of the size for maximum sizing effect. For high free rosin sizes or dispersion sizes the same drying regime is not needed, since these forms of rosin melt and redistribute on fiber surfaces, and react with alumina, at the normal drying temperatures of paper. In this process the moisture content plays a less significant role than it does in the case of the soap size precipitates.

Every mill system represents a different environment. As a result, the rosin can exist in multiple forms when set on fibers by alum and the sintering characteristics of the size precipitates will vary accordingly. The best operating procedures should be worked out on a case-by-case basis under commercial conditions. Find the proper drying regime requires continuing experimentation by the mill personnel.

Application of soap sizes

Maintenance of proper chemical conditions during size emulsification is as important as maintaining the right conditions during the wet-end application. A soap size must be emulsified in soft water and enough alkali must be maintained in the furnish to keep the size in the soap form. This form of size best resists premature agglomeration by polyvalent metal ions.

When sizing with soap size, the common practice is to add the size first and add alum after the size has been uniformly distributed in the paper furnish. It is a good practice to adjust the stock to pH 6.5 to 7.0 prior to the introduction of rosin size. If the pH is between 5.5 and 6.3, the size may undergo premature agglomeration by reacting with polyvalent metal ions, such as Ca++ or recycled alum. Agglomerated size, or size converted to Ca-resinate in hard water areas, is largely lost for sizing. Under these circumstances, it is a common practice to add alum first and then introduce the soap size with good mixing, in order to achieve rapid and uniform size, distribution. If good mixing and rapid addition of alum follows the introduction of size reasonably good results may also be achieved by the standard mode of chemical addition. The pH of the furnish should be controlled before size addition. If the pH is kept too high (about 9.0), the size precipitate contains very little rosin and the sintering point is extremely high and sizing only moderate.

As a general rule the amount of alum required is about 1.5 times the amount of rosin size. For maximum efficiency the alum should be 30% to 50% neutralized, since in this range the alumina possesses maximum cationic charge (28). This amounts to adding 1 OH to 1.5 OH base equivalents per Al+++ ion. The exact amount of base to be added is a function of several factors, such as alum concentration in the furnish, stock temperature and stock pH. These factors determine how much additional OH groups are introduced by hydrolysis. Accordingly, the actual degree of alum neutralization will always be higher than the "formal" degree of neutralization, which is based on the amount of alkali added. The cationic activity, or available cationic charge, has been studied in some detail and the information is published in references 23 and 28. The composition of the most cationic species varies within a narrow pH range, hence pH alone is not a reliable control for maintaining the optimum conditions on machine. The proper base/A ratio must be established by titration. On determining the alkalinity of stock, the base contribution of rosin size must also be taken into consideration.

In highly alkaline systems, excessively large amounts of alum may be required to operate near the optimum pH range. To avoid introduction of unnecessary alum, a mixture of acid (H_2SO_4) and alum is used with beneficial results. Partial replacement of alum with acid in strongly alkaline furnishes results in improved sizing and improved strength. The use of acid also diminishes the amount of latent acidity of the sheet which comes from hydrolysis of the alumina floc, $Al_2(OH)_5(SO_4)_{0.5}$ (21,23) when paper ages. As a result, paper permanency is improved.

Development of sizing is also controlled by monitoring *total acidity*. The total acidity is determined by titrating 100 ml of tray water with 0.02N NaOH to the phenolphthalein end point. Ten times the number of ml NaOH used gives the total acidity. Most mills obtain best sizing when the total acidity is held in the range of 50 to 150 ppm (as $CaCO_3$). When total acidity is too high and/or the pH is too low, an inadequate amount of alumina will be retained and sizing will suffer. If total acidity is too low, either there will be too little alum present or the retained alumina will have insufficient cationic charge. Correlation of total acidities of different mills has shown that the total acidity is not a good index for optimization of alum performance. (69) If correlations between total acidity and alum neutralization are to be drawn, it must be done on a mill-to-mill basis, employing a great deal of historical data. Theoretical work (28) combined with the past (22) and the most recent (69) mill studies provides an unequivocal evidence that the control of the degree of alum neutralization (OH/Al) provides the best approach in optimization of mill performance.

Operation at too low a pH is a source of other problems. The size hydrolyzes to form a colloidal rosin which agglomerates to micellar aggregates and stabilizes foam. Corrosion of equipment is another problem. At low pH, fiber swelling is reduced and paper strength suffers due to reduced fiber bonding. Degradation of mechanical properties and color upon aging become increasingly apparent. Fugitive sizing is another factor, since the free rosin is not adequately bonded, and in the free acid form it becomes increasingly sensitive to auto- oxidation, which reduces sizing.

Continuous addition of size and alum improves process control. Addition of size just before jordaning, and alum after it, is a desirable technique. If size is added early in the system and alum too late, severe foam may result, pitch problems may appear, and sizing may deteriorate because of progressive size agglomeration and poor distribution on fibers.

Application of dispersion sizes

The dispersion sizes represent a significant advance in sizing technology which has afforded the paper industry several benefits: more sizing per pound of rosin used, broader pH range for sizing, improved drainage, improved paper strength, and no need for emulsification equipment. The basis of these benefits lies primarily in the different surface-chemical behavior of dispersion size.

The term "dispersion" implies that the size particles are a solid rosin. The idea of using dispersed rosin for sizing dates back to 1932, when Bewoid size was invented (71). Today, dispersed rosin sizes are offered by several manufacturers, each using a somewhat different process. All these products exceed the performance and handling characteristics of Bewoid size. Dispersion sizes are usually supplied at 40% solids and consist of practically 100% free rosin, which has been to some degree fortified. The dispersions are negatively charged and stabilized using emulsifiers and protective colloids. Mechanical stability of rosin dispersions may vary significantly, and it is not infrequent that a dispersion breaks upon pumping through a high shear device.

Rosin deposits form on equipment parts and the interior walls of pump lines and rotameters. Excessive dilution with water also destabilizes the dispersion, since the effective concentration of the stabilizer is reduced. The dilution water may also contain Ca^{++} or Mg^{++} ions which destabilize the dispersion by reducing its negative charge. Storage stability of a well formulated product is excellent at temperatures from $2°C$ to $40°C$. Temperatures well above this range are harmful and so is freezing of the dispersion, causing a breakdown. Handling procedures that expose the dispersion to shear should be avoided. For instance a relatively stable dispersion can be recycled in a peristaltic pump for more than eight hours, whereas upon recycling through a centrifugal pump or a high shear mixer, its stability is reduced to 450 or 20 minutes, respectively (72).

Size dispersions of small particle size are favored. Typical particle sizes range from 0.1 to 0.5 micrometers. Reaction with alum occurs only with a small fraction of rosin which is in ionized state on the particle surface. Rosin/alum floc formed by a dispersion size is more dense and more compact than the precipitate formed by soap size (70). This is one reason why better drainage is obtained with a dispersion size. Another reason for the drainage improvement lies in the fact that dispersion size does not reduce the cationic charge of alumina (70), while the soap size, by being strongly anionic, does deplete the cationic charge of alumina. The lesser interaction of dispersion size with cationic alumina also has some disadvantages. Dispersion sizes are more difficult to retain when alum is used as the sole retention aid. To achieve maximum effect from a dispersion size, the maintenance of high first-pass retention is a major goal; therefore, cationic polymers, such as wet strength resins and cationic starches, or dual retention aid systems (67) are widely used as adjuncts with alum for setting the size. With cationic polymers the alum dosage can be appreciably reduced, but it should never be undercut below a specific level which is a characteristic of the given furnish. Unbound free rosin migrates (73) and full sizing does not develop.

A suitable procedure is to add alum (with good mixing) at the machine chest and to introduce the dispersion size at stuff box. Trim alum and cationic promoters or retention aids can be added near fan pump or between the fan pump and the machine head box.

In regard to the common mill factors, it is valid to state that the dispersion sizes will be affected approximately the same way as the soap type. For example, the unbleached pulps are easier to size; refining, addition of fillers will increase the size demand; control of the degree of alum neutralization to 30% to 50% significantly benefits size retention and sizing; slimicides or pitch control agents that contaminate sized surfaces may severely hurt sizing of both types. The dispersion sizes do act differently in some respects and this provides certain options to the papermaker. Dispersion size reacts with alumina only when the paper web passes through drier section and aluminum resinate forms even with the neutral aluminum hydroxide. This means that sizing can develop in furnishes where the wet end pH is close to 7. This aspect is discussed in more detail in the following section on "Use of Rosin in Neutral Papermaking".

Summary

The dispersion sizes can cause the papermaker many problems if improperly stabilized or improperly used. If used correctly, the dispersion sizes offer distinct benefits that cannot be matched by the soap size.

Use of rosin in neutral papermaking

With the advent of neutral papermaking which uses $CaCO_3$ as a filler, the question arises if any form of rosin can be used as a sizing agent for these conditions? Contrary to current thinking, the answer is a qualified yes! The effective pH range for sizing with free rosin is not necessarily limited to acid conditions. Technology now exists that allows the extension of the effective range of sizing as high as pH 7.

Operation with the soap size by necessity is limited to acid pH range of 4 to 5. The soap size must be converted to aluminum resinate to be retained in sheet. The reaction is virtually complete in the aqueous phase, before the paper web is formed. If the stock contains $CaCO_3$, the whole system is buffered around pH 7 to 8, and enough alkalinity is present to fully convert the cationic form of alumina to Al $(OH)_3$, which is unable to neutralize the anionic rosin soap. As shown by Figure 1.1, even traces of residual soap convert the hydrophobic surfaces to polar, water-loving and sizing is destroyed.

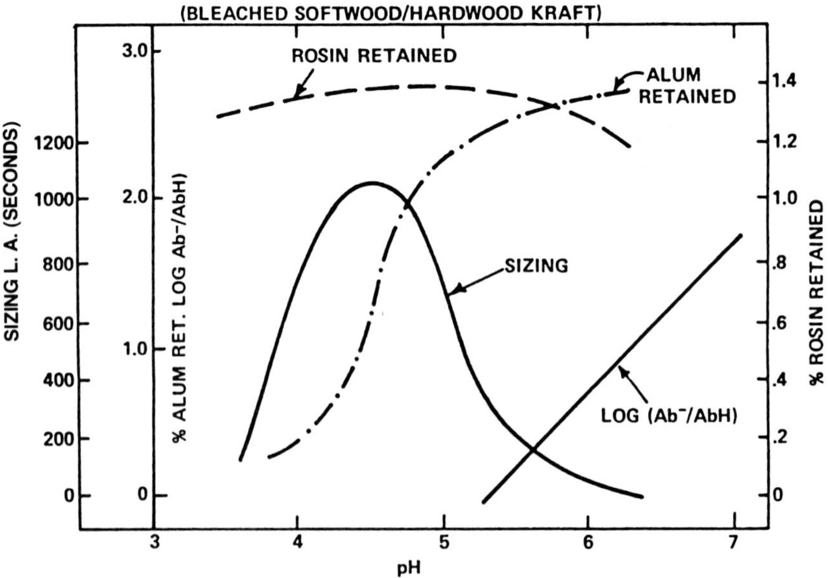

Figure 1.1 Development of sizing as function of pH (bleached softwood/hardwood kraft).

The situation is totally different if the sizing agent is supplied in the form of free acid, such as finely dispersed rosin. Dispersion sizes do not undergo any chemical reaction with alum during retention in the aqueous phase. This form of rosin first reacts with alumina when the paper web passes over the hot drums in the drier section. At this point, high temperature and humidity induce rosin to spread and react with a hydroxy-aluminum polymer on fiber surface, to form aluminum resinate. A continuous water phase no longer exists and the alkalinity of the $CaCO_3$ cannot be passed on to the fiber surface. The result is that a hydrophobic sizing film forms with a minimum contamination from the rosin soap.

This type of chemistry offers the papermaker an option for achieving satisfactory sizing with rosin even in furnishes that contain moderate to high levels of $CaCO_3$ filler. This approach has been implemented on mill scale with considerable success (33). To reduce the extent of interaction of the acidic alum with the alkaline $CaCO_3$ filler, the rosin dispersion and alum are added together, preferably at the thin stock loop. Good sizing has been maintained at a typical operating pH of 6.5, using 0.5% rosin dispersion and 1.0% to 1.5% alum, having a $CaCO_3$ loading of 30%, based on fiber. This approach has enabled the mill to obtain the benefits of $CaCO_3$ filler at modest sizing costs. These benefits include: reduced energy in refining, improved drainage and drying, and the ability to achieve higher filler contents in sheet.

In contemplating the role of rosin in neutral papermaking, one should not lose sight of the fact that in many paper grades most of sizing is accomplished by size-press treatment. For internal sizing just enough rosin size must be used to control the size press pickup. A proper choice of operating procedures in combination with a dispersion size and alum should provide a dependable sizing system for a calcium carbonate filled sheet. The base sheet can be subsequently surface-sized using anionic styrene-maleic acid type copolymers or styrene copolymers with suitable cationic monomers. There are plenty of commercial products already available along these lines, but it is beyond the scope of this chapter to endorse any particular product.

References

1. Kulick, R. J. U.S. Pat. 3,433,658 (1969).
2. Lawrence, R. V. 1962. *Tappi.* 45 (8):654.
3. Wilson, W. S. and Bump, A. H. U.S. Pat. 2,628,918 (1953)
4. Strazdins, E., 1977. *Tappi.* 60 (10):102.
5. Gaddis, V. E. 1962. *Paper Mill News.* Dec. 24, 34.
6. Watkins, S. H. "Rosin and Rosin Size." In *Tappi Monograph Series No. 33.* Atlanta: TAPPI PRESS, 1971.
7. Minor, J.C., Schuller, W.H. and Lawrence, R.V. 1965. *Tappi.* 48 (9):54.
8. Strazdins, E. and Hastings, R., U.S. Patent 2,776,221 (1957).
9. Strazdins, E. 1958. *Tappi.* 41 (10):551.
10. Strazdins, E. and Sheers, E. H. 1958. *Tappi.* 41 (10):658.
11. Swanson, J. W. "The Process of Sizing." In *Tappi Monograph Series No. 33.* Atlanta: TAPPI PRESS, 1971.
12. Strazdins, E. 1981. *Tappi.* 64 (1):31.
13. Strazdins, E. 1965. *Tappi.* 48 (3):157.
14. Back, E., Steenberg, B. 1951. *Svensk Papperstidn.* 54 (15):510.
15. Davison, R. W. 1964. 47 (10):609.

16. Strazdins, E. 1977. *Tappi.* 60 (10):102.
17. Kaltenbach, J. 1954. *Das Papier.* 8 (19/20): 409.
18. Strazdins, E. (1984). *Tappi.* 64 (4):110.
19. Strazdins, E. In *Tappi Seminar Notes on Alkaline Sizing.* Atlanta: TAPPI PRESS, April, 1985.
20. Strazdins, E. 1963. *Tappi.* 46 (7): 432.
21. Reynolds, W. F. and Linke, W. F. 1963. *Tappi.* 46 (7): 410.
22. Reynolds. W. F. 1961. *Tappi.* 44 (2): 177A.
23. Strazdins, E. "Cationic Properties of Alumina." Presented at ACS Symposium on Colloid Science and Surface Chemistry. June 25, 1985, Potsdam, N.Y.
24. Dohne, W. P. and Libby, C. E. 1942. Tech. Assoc. Papers. 25(1):663.
25. Cobb, R. M. K. and Lowe, D. V. 1955. *Tappi.* 38 (2): 49.
25a. Cordier, D. and Bixler, H. J. 1987. *Tappi Journal.* 70 (11): 99.
26. Hinton, R. V. *Tappi Seminar Notes on Retention and Drainage.* Atlanta: TAPPI PRESS, 1982.
27. Ekwall, P. and Bruun, H. 1950. *Paper and Timber* (Finland). 32 (7): 194.
28. Strazdins, E. 1986. *Tappi Journal.* 69 (4): 111.
29. Thomas, A. E. *Colloid Chemistry.*" New York: McGraw-Hill, 1934.
30. Gess, J. W. 1977. *Tappi.* 60 (1): 118.
31. Strachan, J. 1931. *Paper Maker and British Trade Journal.* 82 (4): 86.
32. Thode, E. F., Gorham, J. F. and Atwood, R. H. 1953. *Tappi.* 36(7): 310.
33. Roberts, F. J. and Wilson, C. M. W. 1983. *Pulp & Paper.* September, 1975.
34. Robinson, S. J. 1936. *Paper Trade Journal.* 103(7): 130.
35. Verhoeff, J., Hart, J. A. and Gallay, W. 1963. *Pulp Paper Mag. Can.* 64, T 509.
36. Ecke, A. 1933. *Papierfabrikant.* 31 (50):667.
37. Mispley, R. G. 1964. *Tappi.* 47 (1):58A.
38. Puzyrev, S. S., Bozhick, B. and Ivanov, S. N. 1971. *Bum, Prom.* (3), 11.
39. Edge, S. R. H. 1930. *World's Paper Trade Rev.* 92(22): 1974.
40. Wilson, W. S. 1949. *Tappi.* 32 (9):429.
41. Marton, J. and Marton, T. 1982. *Tappi Journal.* (11):105.
42. Eklund, D. 1969. *Paper and Timber* (Finland). 51(4A): 251.
43. Eklund, D. 1967. *Nor. Skogind.* 21(4):140.
44. Sizing Technical Highlights by American Cyanamid Co. p. 81. 1963.
45. Strazdins, E. 1972. *Tappi.* 55 (12): 1691.
46. Volkov, V. A. and Yurev, V. I. 1974. *Bum. Prom.* (3), 9.
47. Lindstrom, T. "Adsorption of Wet End Additives Onto the Surface of Pulps and Fillers." In *Tappi Notes on Advanced Topics in Wet End Chemistry Seminar.* Atlanta: TAPPI PRESS, 1985.
48. Brooks, A. M. 1958. *Tappi Monograph No. 19.* p. 10.
49. Brecht, W. and Heyn, D. 1966. *Das Papier.* 29 (5):238.

50. Reynolds, W. F., et al. 1957. *Tappi.* 40 (10): 839.
51. Casey, J. P. 1942. *Tech. Assoc. Papers.* 25:139.
52. Casey, J. P. 1942. *Tappi.* 37 (6): 152A.
53. Weber. C. G. and Shaw, M. B. 1935. *Nat. Bur. Stand,* (U.S.A.) Misc. Publ. 150.
54. Chilson, W. A., et al. 1934. *Tech. Assoc. Papers.* 17 (1): 381.
55. Davidson, P. B. 1954. *Tappi.* 37 (1): 18.
56. Weiger, B. 1935. *Paper Trade Journal.* 101 (5):51.
57. Rowland, B. W. and Bain, W. M. 1940. *Paper Trade Journal.* 110 (17):237.
58. Sutermeister, E. 1924. *Am. Dyest. Rep.* 13, 515, 548.
59. Tucker, E. C. 1923. *Paper Trade Journal.* 76 (18): 47.
60. Hammill, G. K. et. al. 1927. *Paper Trade Journal.* 84 (3): 39.
61. Collins, T. T., Davis, H. I. and Rowland, B. W. 1942. *Tech. Assoc. Papers.* 25 (1): 606.
62. Harrison, H. A. 1932. *Paper Maker Br. Paper Trade J.* 83, TS 158.
63. Bridge, F. and Harrison, H. A. 1947. *Proc. Tech. Sec. Paper Makers Assoc. of Great Britain and Ireland.* 28 (1): 249.
64. Klemm, P. 1908. *Wochenbl. Papierfabr.* 89 (18) 1367.
65. Arledter, F. 1907. *Papier Ztg.* 32 (18) 773.
66. Strazdins, E. 1980. *Das Papier.* 34 (10A) 49V.
67. Strazdins, E. 1984. *Pulp & Paper.* 58 (3) 73.
68. Strazdins, E. "New Insights Into Aluminum Chemistry Related to Papermaking. Presented at XXII EUCEPA Conference, October 6-10, 1986, Florence, Italy.
69. Cordier, D. R. and Bixler, H. J. "Measurement of Aluminum Dydrolysis in the Wet End (to be published in *Tappi Journal* in 1987).
70. Kulick, R. J. 1977. *Tappi.* 60 (10): 74.
71. Wieger, B., U.S. Pat. 1,882,680 (1932).
72. Alphasize 20 Product Bulletin by American Cyanamid Company.
73. Gess, J. M. "Rosin Sizing: A New Approach." In *Tappi Papermakers Conference Proceedings.* 9-14. Atlanta: TAPPI PRESS, 1982.

2

Alkyl Ketene Dimer Sizes

R. E. Cates, D. H. Dumas, and D. B. Evans

Introduction

Alkyl ketene dimer (AKD) has been used as a sizing agent for more than 30 years (1). It is still considered by many to be something of a specialty with applications directed toward the longevity of paper for archival purposes. In fact, ketene dimer sizes are used in virtually every segment of the paper industry; from newsprint to fine paper, from lightweight coated to gypsum board.

There are two reasons that may account for the misconceptions about ketene dimer size. One, the size is cellulose reactive and thus does not function by the same process as does typical rosin size. Two, the size works in an alkaline pH range allowing paper to be made without alum under papermaking conditions quite different from normal conditions. The characteristics of alkaline papermaking (often an important part of the driving force for use of AKD) are examined later in this chapter. AKD properties, sizing mechanism, and conditions of use are discussed below.

Chemical properties

Ketene dimers can be represented by the structure shown in Figure 2.1 (8):

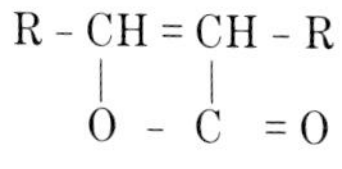

$$R - CH = CH - R$$
$$O - C = O$$

Figure 2.1 Structure of ketene dimer.

AKD for paper sizing is prepared by the dimerization of aliphatic acid chlorides prepared from fatty acids. The choice of fatty acid influences the melting point of the AKD. Unsaturated fatty acids lead to a liquid AKD, whereas saturated fatty acids give solid products. AKD is typically made from commercial stearic acid (R = C_{14}–C_{16} in Figure 2.1) and has a melting point of about 50°C.

Sizing efficiency increases with carbon number from C_8 to C_{14} and levels off with only minor variation up to C_{20} (see Table 2.1). There are literature reports suggesting that sizing should increase with increasing carbon number because of the wider arc swept by the longer fatty acid chain (2). This does not seem to be the case with AKD above C_{14}, possibly because of the two chains present in each molecule. The chains together can sweep out a larger area, which probably minimizes the effect of chain length.

Table 2.1 Effect of chain length on AKD sizing efficiency (65 g/m² handsheets with 0.1% size added).

Chain Length*	HST to 80% Reflectance
Mixed C_{14} - C_{16}	786 sec
Pure C_{16}	825 sec
Mixed C_{18} - C_{20}	700 sec
Pure C_{20}	675 sec

*Chain length refers to R group as given in Figure 2.1. Chain length of fatty acid used to make AKD is equal to R + 2 carbons.

Sizing mechanism

Paper is sized to provide resistance to penetration by liquids, typically water or aqueous solutions. Sizes can be added "internally" to the pulp slurry at the wet end of the paper machine, or "surface applied" to the dried sheet at the size press or calender stack. Internal sizes function by making the fiber surfaces in the sheet more hydrophobic. This increases the contact angle formed between the fibers and the penetrating liquid, thereby decreasing the capillary penetration rate. Surface sizes can also function by increasing the surface hydrophobicity, or by film formation (i.e., reduction of pore size in the sheet surface). As with most other commercial sizes, AKD is used primarily as an internal additive and functions by increasing the fiber surface hydrophobicity.

All internal sizing agents must undergo the following four processes to provide effective liquid repellency:

1. Retention on fibers in the forming mat
2. Distribution over the fiber surfaces
3. Orientation on the fiber surfaces to present a hydrophobic surface to the penetrant
4. Anchoring to prevent overturn by the penetrant.

The relative efficiency of any given size can be explained on the basis of its ability to achieve these four objectives (3). The high efficiency of AKD, which can give significant sizing at 0.02 - 0.03% added AKD (4), can be explained on this basis.

AKD retention

There is a direct relationship between AKD wet-end retention and sizing as shown in Figure 2.2. AKD is effectively retained using cationic starch or cationic resins (5,6) or by use of cationic AKD emulsions (7). The retention occurs by charge attraction between the anionic cellulose and the cationic AKD particles. To maximize the retention on fiber, the size is usually added close to the headbox, typically between the stuffbox and screens. Both high shear and long contact time are known to reduce AKD retention (5).

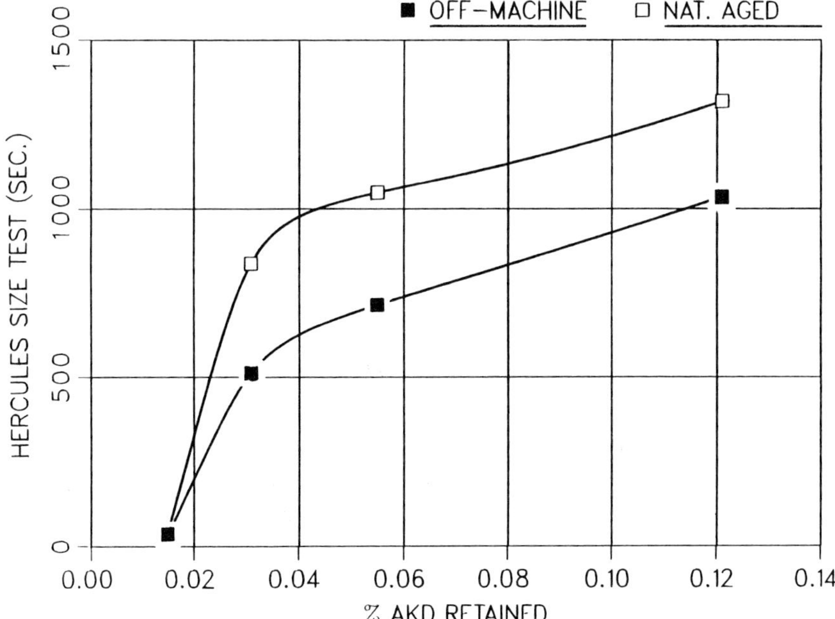

Figure 2.2 AKD retention vs. sizing.

The size particles also adsorb on fillers and fines in the wet end. Since fines and fillers adsorb a disproportionate amount of size due to their high surface area, increasing filler retention usually also increases sizing efficiency. This effect is shown in Table 2.2. These results were obtained using a cationic AKD emulsion with cationic starch and a high- molecular-weight retention aid. When a high-molecular-weight anionic retention aid is post- added after resin ("dual polymer" retention system), AKD retention of over 90% is possible

(4,5). Thus, good AKD retention can be obtained using either special cationic emulsions or via typical additives used to retain fines and fillers.

Table 2.2 Effect of flocculants on filler retention and sizing (65 g/m² Pilot machine-made paper containing 10% calcium carbonate and 0.25% cationic starch).

| | % Retention | | HST to 80% |
Flocculant	Overall 1st Pass	Filler	Sizing
None	89%	56%	72 sec
0.025% Cationic Polyacrylamide	93%	67%	95 sec
0.025% Anionic Polyacrylamide	95%	77%	101 sec

AKD distribution

The distribution of AKD size on the fibers occurs in the dryer section of the paper machine. Here, heat from the dryers melts the retained AKD particles and spreads them over the available fiber surfaces. The low melting point of AKD allows it to be spread much more efficiently during drying than the alum-rosin precipitate, which has a relatively high sintering temperature.

AKD cellulose reaction

The ability of AKD to be oriented and immobilized on fiber surfaces is related to its molecular structure and reactivity. The four-membered lactone ring in AKD allows it to react with nucleophiles to form β-keto derivatives. This is illustrated in Figure 2.3 for the reaction between AKD and cellulose.

The reaction of AKD with cellulose to form the β-keto ester is the primary mechanism by which AKD sizes paper. This covalent bond formation provides immobilization and orientation of the hydrophobic tails outward, away from the fiber surfaces. In addition, this anchoring/orientation process is very efficient since it occurs on a molecular scale.

It should be noted that there has been a recent controversy in the literature over the actual mechanism of AKD sizing. The AKD-cellulose reaction mechanism has been traditionally supported by the efficient sizing of AKD, and the good resistance obtained to both acid and alkaline penetrants. In addition, paper sizing is unaffected when a solvent for AKD (e.g., tetrahydrofuran or chloroform) is used to extract an AKD sized sheet.

Recently, AKD-cellulose reactivity has been questioned based on the inability to analytically detect ester bonds in AKD-sized paper or with cellulose model compounds (11, 12, 13). This has led to alternative explanations of

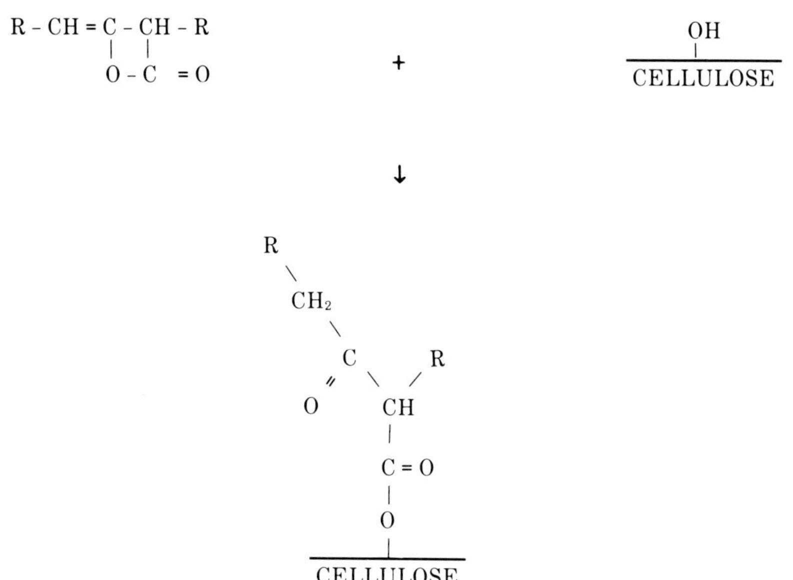

Figure 2.3 AKD - cellulose reaction.

AKD sizing that do not require ester formation (14, 15). However, these alternative theories are severely weakened by several new observations. A series of studies using high-pressure liquid chromatography (16) and C_{14}-labelled AKD (4,9,17) has demonstrated conclusively that sizing is provided by an unextractable fraction of the retained AKD in paper. In one study, it was established that the AKD-cellulose reaction follows pseudo first-order kinetics (6). Also, direct spectroscopic evidence for the formation of keto-ester bonds has been demonstrated for AKD with cellulose and cellulose model compounds (18).

AKD-water reaction

AKD can also react with water as shown in Figure 2.4. This hydrolysis reaction forms the unstable β-keto acid which decarboxylates to form a ketone. The reaction between AKD and water at room temperature is slow enough that commercial dispersions of AKD in water remain stable for at least a month. Also, very little AKD hydrolysis occurs during drying in typical papermaking systems (4,9,19). The hydrolysis product of commercial AKD size, stearone, has been traditionally viewed as a non-size (10). However, recent work shows that it may actually contribute to sizing when in the presence of reacted AKD (9). Further information on stability and hydrolysis of commercial AKD sizes is given later in the chapter.

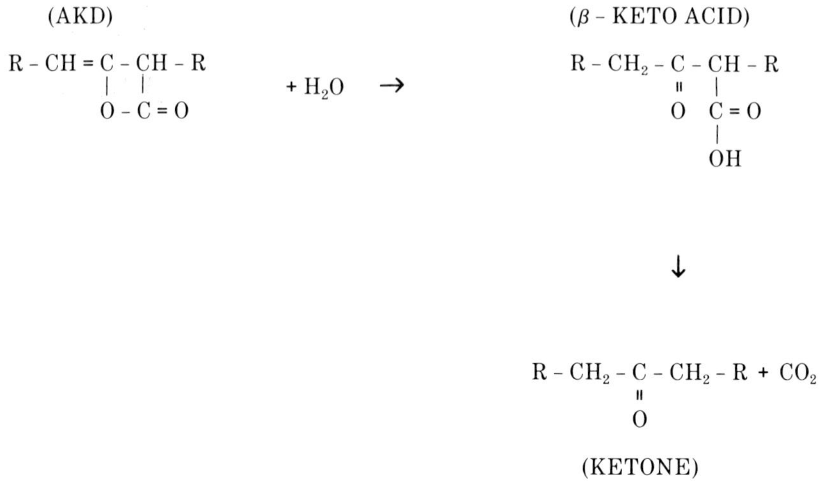

(AKD)

$$R - CH = C - CH - R$$
$$\quad\quad\; | \quad |$$
$$\quad\quad O - C = O$$

$+ H_2O \quad \rightarrow$

(β – KETO ACID)

$$R - CH_2 - C - CH - R$$
$$\quad\quad\quad\; \| \quad\; |$$
$$\quad\quad\quad O \quad C = O$$
$$\quad\quad\quad\quad\quad |$$
$$\quad\quad\quad\quad\quad OH$$

$\downarrow$

$$R - CH_2 - C - CH_2 - R + CO_2$$
$$\quad\quad\quad\; \|$$
$$\quad\quad\quad\; O$$

(KETONE)

Figure 2.4 AKD - water reaction.

Emulsion properties

AKD sizes are usually sold in the form of aqueous emulsions which range in properties from slightly cationic to highly cationic. The solids content of the emulsions ranges from 6% to about 15%.

Since alkyl ketene dimer can react with water as well as cellulose, the shelf life of the commercial emulsions is limited. Shelf life of commercial AKD emulsions at various temperatures is shown in Table 2.3.

Table 2.3 Shelf life of commercial AKD emulsions.

Storage Temperature	Recommended Maximum Shelf Life
32°C	2 weeks
24°C	4 weeks
4°C	1 year

AKD sizes are shipped in drums or in bulk and handled in conventional storage and metering systems. The products are usually metered to the paper machine in concentrated form, and diluted with water at the point of addition to improve distribution on the fiber. They are added close to the headbox to avoid prolonged contact with the stock. As discussed previously, retention conditions and additives which provide maximum retention of fines and filler will usually give good retention of AKD size.

Practical considerations

Successful use of AKD in paper requires knowledge of the effect of a number of papermaking variables on AKD sizing. Some of these effects occur through known changes in AKD retention or reactivity, while others have simply been observed through many years experience in the field. Some of the more important papermaking variables are discussed below along with suggested procedures for optimizing AKD use.

Effect on pH, alkalinity, and alum

The effects of alum levels, pH, and alkalinity on AKD sizing are interrelated. The pH of the papermaking system can affect sizing, but it is not independent of the alkalinity. Figure 2.5 shows the results from a pilot machine trial with 100 ppm of dissolved alkalinity in a clay filled sheet. The sizing is measured three times for each condition:

1) Before the size press (at 3% to 5% moisture)
2) At the reel (immediately off the reel at 4% to 6% moisture)
3) After one week at 50% R.H. and 22°C.

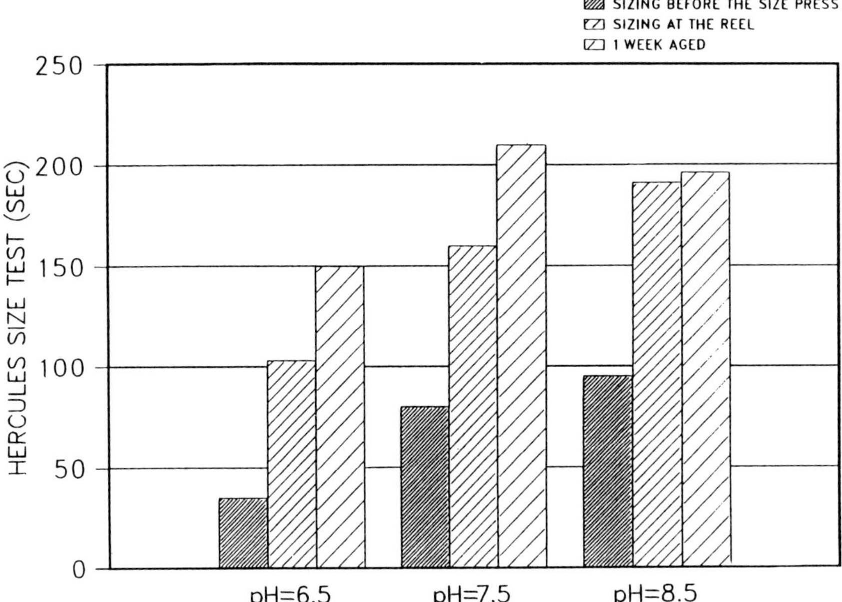

Figure 2.5 Effect of pH on the rate of size development.

As the pH increases from 6.5 to 7.5, there is an overall increase in the sizing. As the pH further increases to 8.5, the rate of sizing continues to increase, as seen before the size press, but the ultimate degree of sizing levels off (see 1 week aged data). AKD sizes have generally been found to work effectively over the pH range from 6 to 9.

Alkalinity is a major contributor to the effective reaction between AKD and cellulose. This is shown graphically in Figure 2.6. Increasing the alkalinity from less than 10 ppm to 150 ppm, expressed as $CaCO_3$, results in dramatic increases in sizing both off-machine and natural aged. With sufficient bicarbonate alkalinity, low levels of alum can be tolerated with only a minor effect on the natural aged sizing. Use of excess alkalinity (Table 2.4) has no detrimental effect on sizing. Based on these results and many years experience in the field, 150 to 250 ppm dissolved alkalinity is recommended. Use of calcium carbonate fillers, which act as buffers between pH 7.5 to 8.5, is beneficial for AKD sizing. The pH is stabilized and a source of alkalinity is provided. However, the solubility of calcium carbonate is very low. Unless there is ample contact time between the $CaCO_3$ and the stock, it may be necessary to add soluble alkalinity with soda ash or bicarbonate.

The ability of bicarbonate ion to increase the reactivity of cellulose with AKD is not well understood. Since alkalinity provides faster, more complete reaction analogous to increasing the pH, it suggests that the cellulose itself is affected. One explanation is that high alkalinity results in a more reactive,

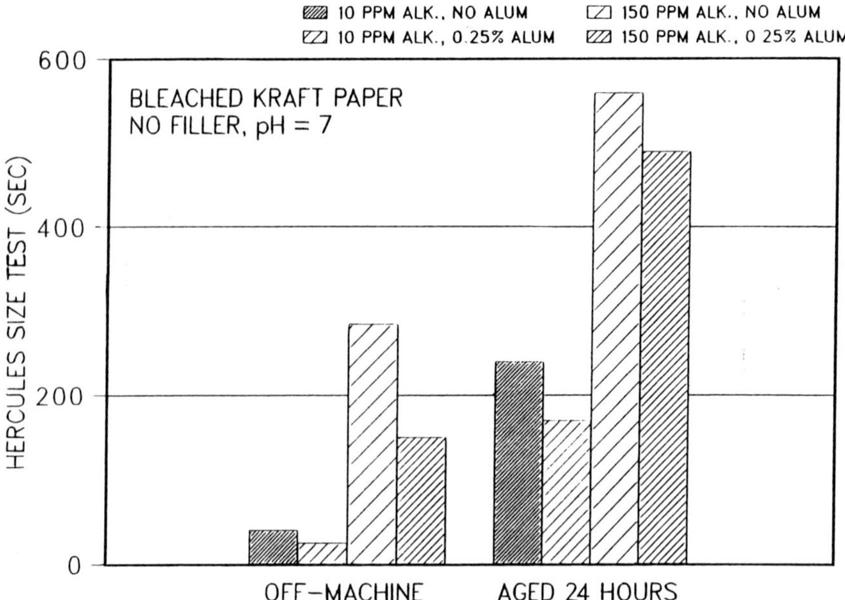

Figure 2.6 Effect of high alkalinity on rate of size development.

Table 2.4 Effect of high alkalinity on AKD sizing.

	Hercules Size Test of 80% Reflectance (sec)	
ppm Alkalinity	*Off Machine*	*1 Week Aged*
150	180	264
250	177	251
550	183	247
1000	154	239

less internally hydrogen bonded hydroxyl group, anda more swollen pulp surface. Increasing pH certainly causes the cellulose to swell and become more reactive. Another explanation involves the bicarbonate ion as a proton transfer catalyst from AKD to cellulose (6,9).

Regardless of the mechanism, addition of soluble alkalinity is the easiest way to ensure a stable system pH, to increase the sizing efficiency of AKD sizes, and to counteract the negative effect of alum. There are usually high levels of alkalinity in both calcium carbonate filled and unbleached kraft systems. In these systems, addition of alum can actually increase sizing because of its value as a retention/drainage aid.

Effect of fillers

The addition of filler to a pulp system results in increased demand for size to maintain the same level of sizing. There are two reasons for this increased size demand:

a) Increased surface area of the fillers which must be sized.
b) Lower retention of the filler relative to the fiber.

When the size is added to the pulp/filler slurry, it adsorbs on the surfaces of the filler and pulpin relation to their respective surface areas. This means that a disproportionate amount of size will adsorb on the filler. Depending on the type, filler has from 2 to 20 times as much surface area as pulp. The filler is then retained in lower amounts than the pulp due to the small size of the filler particles. This observation is true for rosin based sizes as well.

With an alkaline size, the papermaker now has the opportunity to use calcium carbonate filler. There are two types of calcium carbonate available today, precipitated and finely ground. The differences between these two types are manifold, but one which is important for this discussion is the average particle size. Typical precipitated products are uniform in size and about 0.5 μm in diameter. Filler grade (fine ground) calcium carbonate is about 2 μm in diameter and somewhat less uniform in size than the precipitated products. The effect of filler type and loading level on sizing and tensile strength is shown in Figures 2.7 and 2.8.

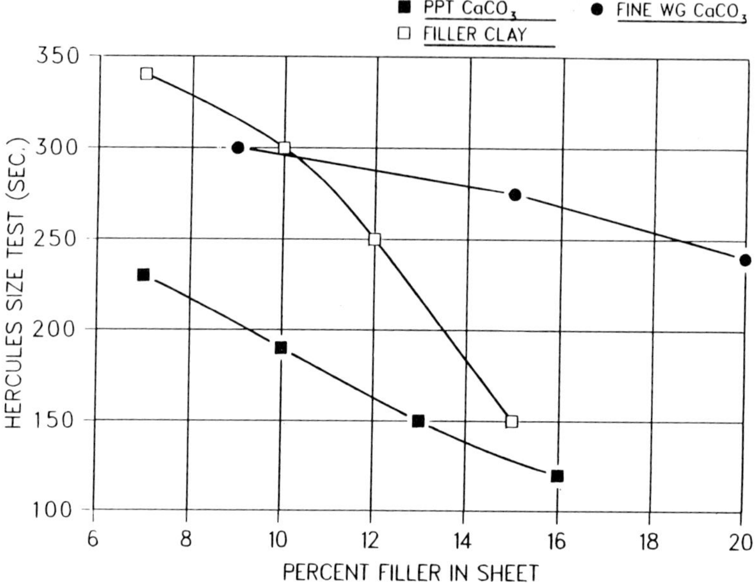

Figure 2.7 Sizing vs. loading.

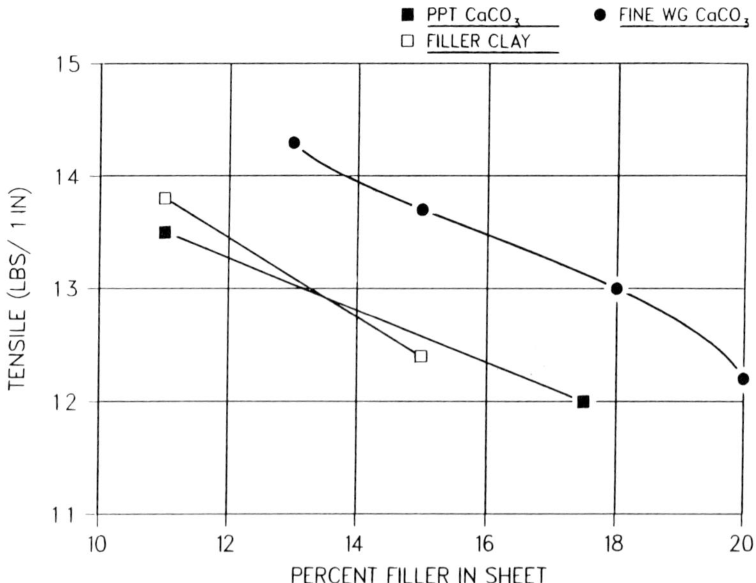

Figure 2.8 Tensile vs. loading.

The losses in sizing and tensile strength with added filler are directly related to surface area (particle size) effects. Because of this, fine ground carbonate offers significant advantages in this area. Since an alkaline pH and dissolved alkalinity are desirable for AKD sizing and since higher filler loadings and brightness are desirable for papermaking, calcium carbonate is an excellent filler to use with AKD. Further benefits associated with calcium carbonate use are described later in the chapter.

Pulp type

Pulps differ in their reactivity and sizeability with respect to AKD. Some pulps react more slowly and to a lower extent, while others react effectively and size more easily. The surface area of the pulps is another factor that affects sizing. Examples of these effects are shown in Table 2.5 where the percent reacted AKD (of that retained), the amount reacted, and the resulting sizing are shown. The kraft pulps show similar reactivity with AKD, but the unbleached kraft sizes much more easily. This can be explained at least in part by surface area differences. The bleached sulfite is somewhat less reactive. The acid sulfite pulping process produces a pulp with less reactivity than the alkaline kraft process. In bleached pulps, the differences between kraft and sulfite are reduced because of the alkaline extraction stages which remove the residual acidity. The rag pulp is the least reactive due to its high alpha content; however, it is also difficult to size. It requires almost 10 times as much reacted AKD to provide sizing similar to that of kraft pulps. The reason is unknown, but surface area is not a viable explanation.

Table 2.5 Effect of pulp type on reactivity and sizing efficiency (65 m²/g handsheets).

Pulp Type	% Reacted	Amount Reacted	HST to 85% Reflectance (sec)
Bleached Softwood Kraft	65%	0.028%	530
Unbleached Kraft	67%	0.012%	2700
Bleached Sulfite	54%	0.039%	850
Rag	43%	0.123%	1010

It is also known that mechanical pulps such as TMP or waste groundwood are quite difficult to size. In these systems, the cationic AKD emulsions have been found to be particularly effective (20).

Other additives

Anionic materials such as dispersants or defoamers can interfere with sizing. They prevent effective retention and they detract from sizing if they are retained in the sheet. Small amounts of alum or strongly cationic retention aids are often used to neutralize these anionic species.

Alkaline papermaking using AKD

Alkaline sizing is the term that has been used to describe the production of paper using a cellulose reactive size. In recent years, the paper industry has come to use the term alkaline papermaking to describe the entire process of converting from conventional acid sizing to alkaline sizing. The benefits of alkaline papermaking have resulted in a large number of paper and board machines throughout the world converting to alkaline sizing using alkyl ketene dimer size (21,22,23). These benefits are described in more detail in the following sections.

Use of AKD for highly-sized paper and board

Two critical requirements for the production of sized paper are that the sizing material have a strong adhesion to the cellulose fiber, and that the size not be soluble in the penetrant. The covalent bond produced between AKD and cellulose provides the strongest possible adhesion to the fiber. The new chemical compound formed by reaction of the AKD with the cellulose is insoluble in virtually all aqueous penetrants and organic solvents. The permanence of the AKD-cellulose bond also permits long-term exposure to liquids without loss of sizing. These sizing properties are used to obtain:

> Resistance to strongly acidic or alkaline penetrants.
> Resistance to unusual penetrants containing solvents or surfactants.
> Resistance on long-term exposure to strong penetrants.

AKD size is widely used in grades requiring hard sizing and durability, such as liquid packaging board, meat wrapping paper, and photographic paper.

Paper produced under neutral or alkaline conditions is stronger than paper produced under acid conditions with alum. Alkaline paper has both higher tensile strength and higher elongation than acid paper. Also, in many grades of paper and board, up to 1% to 2% rosin and 1% to 2% alum can be replaced with as little as 0.1% to 0.2% AKD size. The much smaller amount of hydrophobic material results in less interference with fiber bonding in the finished paper.

Use of AKD in calcium carbonate filled papers

Calcium carbonate is a highly desirable filler in papermaking because of its high brightness, low cost and general availability. However, since it dissolves under acid conditions, it cannot be used on paper machines operating below about pH 6.5. Alkaline papermaking using AKD size has made it possible to obtain significant advantages by using calcium carbonate in filled grades (24,25,26). These include:

> 5% to 10% higher filler loading with no loss of productivity or sheet properties.
>
> Partial substitution of calcium carbonate for titanium dioxide.

In addition, significant gains in paper optical properties can be obtained by using calcium carbonate instead of clay. This is illustrated in Figures 2.9 and 2.10 which show the effect of filler type on paper opacity and brightness. Both the precipitated and ground $CaCO_3$ give higher brightness than clay, while the precipitated product also gives superior opacity.

Alkaline papermaking using calcium carbonate filler and AKD size is being used in a number of printing and writing and xerographic grade mills (7).

The inherently higher strength of alkaline paper and board, coupled with the elimination of alum and the use of only small amounts of AKD size, have resulted in significant productivity advantages in mills which have converted to alkaline papermaking. Specific advantages that have been reported are:

> Better machine drainage due to lower additives level
> Improved drying due to lower additives level
> Less refining required
> Reduced alum and additive deposits
> Less corrosion

Alkaline papermaking systems using AKD run relatively clean and require few washups. The rate of AKD size hydrolysis at wet end conditions is slow, and the small amount of solid, high-melting hydrolyzate formed is readily retained in the paper.

Productivity benefits are obtained on all types of alkaline paper and board. Obtaining the full benefits of alkaline papermaking may be a slow process, because of unfamiliarity of the operators with operation of a machine under alkaline conditions. It may require up to 6 months after conversion for a mill to achieve all of the potential productivity benefits such as those listed above.

Use of AKD for increased white water recycle

The use of alum in acidic sizing systems usually significantly reduces the ability of the mill to close up its white water system. In the case of integrated pulp and paper mills, the acidic white water contains soluble aluminum compounds

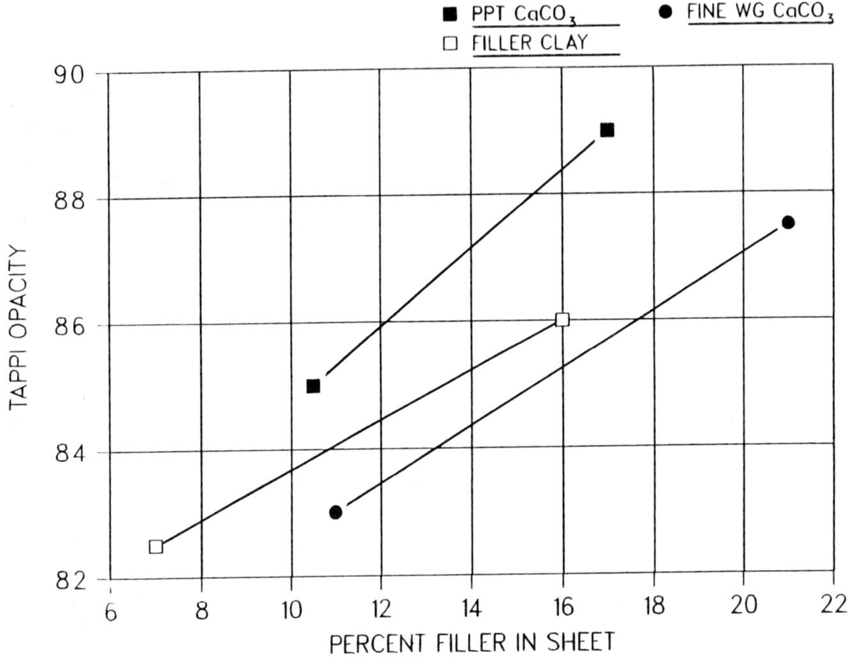

Figure 2.9 Opacity vs. loading.

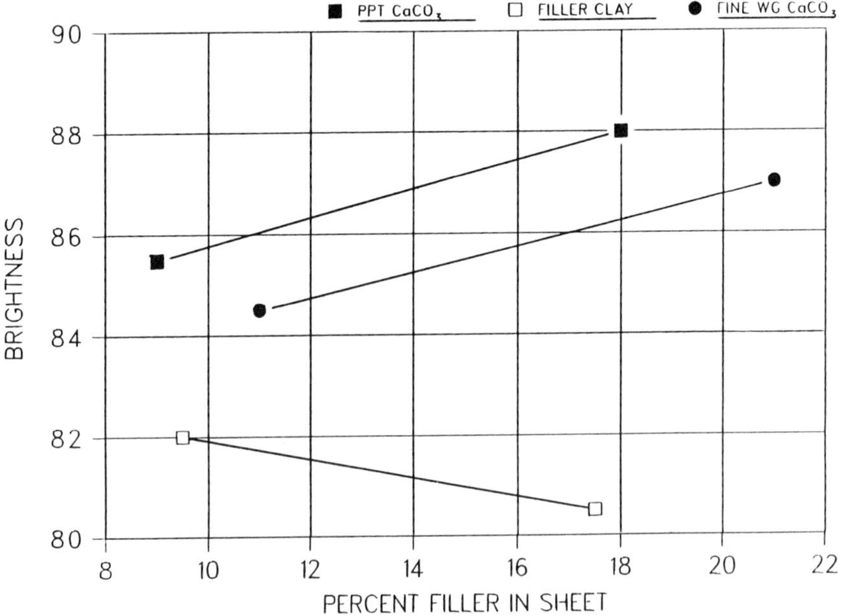

Figure 2.10 Brightness vs. loading.

and is not suitable for use in many parts of the pulp mill or bleach plant. In non-integrated mills using alum, the buildup of acidity, sulfate ion, and dissolved inorganic salts can result in corrosion of the stock preparation system and paper machine.

Cylinder board mills and other mills using recycled pulp are often located in or near large cities and have problems with both incoming water availability and cost, and discharge of waste water. A number of these types of mills have now closed their white water systems completely by conversion to alkaline papermaking and the use of AKD size.

Use of AKD for improved paper performance

Paper and board produced at alkaline pH is more resistant to strength loss on aging than acidic papers. When calcium carbonate filler is used, the paper is buffered to resist acidification by the absorption of carbon dioxide or sulfur dioxide from the air, and permanence is further increased (27,28). AKD sizing is used, usually with calcium carbonate fillers, in archival type papers for records and for permanent book papers.

Potential problems in alkaline papermaking using AKD size

Because alkaline papermaking is different from acidic papermaking, paper machine operators and managerial personnel have to learn to cope with somewhat different problems and to use different solutions in order to obtain the benefits of alkaline papermaking using AKD size (29,30). Problems that have been reported and their solutions are listed below.

Slime formation - Many bactericides that are highly effective under acidic conditions are ineffective at neutral to alkaline pH. By converting to different types of commercial bactericides, or by using chlorine or chlorine dioxide, mills running alkaline using AKD size are controlling slime economically (31).

Fast drainage and poor formation - Due to elimination of large amounts of wet-end additives, the alkaline sheet drains more rapidly on the wire than an acid sheet. Too rapid removal of water from the wire causes poor formation. By learning about the different appearances of an alkaline sheet on the wire and by changing foils and foil box vacuums, papermakers achieve normal drainage and good formation when running alkaline (32).

Increased wear on fabrics - Machines running high levels of calcium carbonate initially noted increased fabric wear. This was solved in part by controlling drainage so that the flat boxes are not run dry. Use of ceramic foil covers and ceramic flat box covers also minimizes fabric wear in paper with high carbonate loading (32).

Wet press picking - Wet press picking has been noted primarily in the early stages of conversion to AKD size and alkaline papermaking. Picking is usually not a problem after rosin-containing stock and broke are out of the system, and after wire drainage and first pass retention are optimized (33).

Insufficient sizing ahead of the size press - For most paper and board machines, only a moderate level of sizing development is required for runnability at the size press. By optimizing wet-end pH and alkalinity, or by using cationic AKD sizes which have a higher cure rate, mills using AKD sizes are achieving good size press runnability on all types of paper and board (2,20).

Paper slipperiness - Alkaline sized paper and board tends to be more slippery than rosin- sized grades. In filled paper grades the substitution of at least part of the clay with calcium carbonate results in good friction characteristics. This is due to the irregular particle shape of the calcium carbonate. In unfilled grades, the use of calcium carbonate in size press solutions, or use of anti-slip agents at the calender stack, controls paper and board slipperiness (34).

Difference in alkaline and acid paper properties - Some differences in alkaline paper properties are so slight as to be almost subjective. Others may be due, for example, to increased filler content, or the fact that an alkaline sheet is stronger. These problems are usually solved by education of customers concerning alkaline paper and board properties, and by making minor changes in paper specifications.

Surface sizing using AKD

AKD size may also be used in surface sizing applications at the size press or calender stack, or in off-machine operations. AKD size is highly efficient when applied to the surface, since retention is 100% and water is removed from the sheet rapidly, giving maximum opportunity for reaction with the cellulose. AKD levels used in surface sizing applications range from 0.01% to about 0.05%.

Surface sizing solutions containing AKD size must be kept at as low a temperature as possible to prevent reaction of the AKD with the water. Solution temperature should preferably be below 66°C and storage time should be less than two hours. Where temperature or storage time exceed these parameters, the AKD size may be metered continuously to the surface sizing solution supplying the size press or calender, and recycle of the solution should be kept to a minimum (20).

AKD emulsion may be used with starch or other surface additives, but compatibility at the concentration and temperature of the solution should always be checked. AKD size must never be cooked with starch because the high temperature of the cooking process will completely destroy its sizing efficiency.

As with surface applications of wax size, surface application of AKD size may result in increased slipperiness of paper. AKD addition at the size press should therefore be kept to a minimum. Use of calcium carbonate in size press solutions, or use of chemical anti-slip agents at the calender stack will usually control slip properties.

References

1. Davis, J. W., Roberson, W. H. and Weisgerber, C. A. 1956. *Tappi.* 39 (1):21.
2. Swanson, R. E. 1978. *Tappi* 61(7):77.
3. Davison, R. W. "A General Mechanism for Internal Sizing." In *1986 TAPPI Papermakers Conference Proceedings.* 17-28. Atlanta: TAPPI PRESS, 1986.
4. Dumas, D. H. and Evans, D. B. "AKD-Cellulose Reactivity in Papermaking Systems." In *1986 TAPPI Papermakers Conference Proceedings.* 31-35. Atlanta: TAPPI PRESS, 1986.
5. Davison, R. W. "Retention of Alkyl Ketene Dimer Size in Papermaking Systems." In *1985 TAPPI Papermakers Conference Proceedings.* 7-16. Atlanta: TAPPI PRESS, 1985.
6. Lindstrom, T. "On the Mechanism of Sizing with Alkyl Ketene Dimers." Paper presented at XXI Eucepa International Conference. Torremolinos, Spain. 1984.
7. Johnson, R. G. "Alkaline Printing and Writing Production Using Alkyl Ketene Dimer Sizes." In *1985 TAPPI Alkaline Papermaking Seminar Notes.* 85-88. Atlanta: TAPPI PRESS, 1985.
8. Hanford, W. E. and Sauer, J. C. *Organic Reactions.* Edited by Adams, R., et al. 108-140. Wiley, N.Y., 1946.
9. Lindstrom, T. "On the Mechanism of Sizing with Alkyl Ketene Dimers." In *1986 TAPPI Papermakers Conference Proceedings.* 31-35. Atlanta: TAPPI PRESS, 1986.
10. Dumas, D. H. 1981. Tappi. 64 (1):43.
11. Enke, M. 1973. *Cellulose Chem. Tech.* 7:487.
12. Pisa, L. and Murckova, E. 1981. *Papir Celuloza.* 36 (2): V16.
13. Roberts, J. C. and Garner, D. N. 1984. *Cell. Chem. Tech.* 18:275.
14. Rohringer, P., Bernheim, M. and Wertheman, D. 1985. *Tappi Journal.* 68 (1):83.
15. Lund, R. C. "Studies on the Mechanism of AKD Sizing." In *1985 TAPPI Alkaline Papermaking Conference Proceedings.* 1-5. Atlanta: TAPPI PRESS, 1985.
16. Kamutzki, W. and Krause, T. 1982. *Das Papier.* 36:311.
17. Roberts, J. C. and Garner, D. N. 1985. *Tappi Journal.* 68 (4):118.
18. Nahm, S. H. 1986. *J. Wood Chem. Tech.* 6 (1):89.
19. Kamutzi, W. and Krause, T. 1983. *Wochenbl. Papierfabr.* 7:215.
20. Alberts, R. M. "Alkyl Ketene Dimer Sizes." *TAPPI Sizing Seminar Notes.* 111-115. 1986.
21. Matters, J. F. 1983. *Southern Pulp & Paper.* 46 (12):30.
22. Bryson, H. R. July 1983. *Pulp & Paper.* 57(7):74.
23. Colgan, H. P. "Review of Experience With the Alkaline Process in the U.S.A." Presented at the *Third International Seminar on Paper Mill Chemistry.* Boston, Mass. 1981.
24. Evans, J. C. W. 1983. *Pulp & Paper.* 47(5):100.

25. Anderson, T. C. "Use of Precipitated Calcium Carbonate in Alkaline Systems." In *TAPPI Alkaline Papermaking Conference Notes*. 27-23. Atlanta: TAPPI PRESS, 1983.

26. Adams, F. R. Ground Calcium Carbonate in Alkaline Papermaking." In *TAPPI Alkaline Papermaking Conference Notes*. 1-6. Atlanta: TAPPI PRESS, 1983.

27. Pickwood, N. 1981. *Paper*. 195(8):29.

28. Bryson, H. R. 1981. *Paper*. 195(11):48.

29. Bryson, H. R. "Neutral/Alkaline Papermaking: A Review of Some Potential Benefits." Presented at the *Third International Seminar on Paper Mill Chemistry*. Boston, Mass.

30. Atkinson, J. G. 1982. *Paper Trade Journal*. 166(3):30.

31. Goldstein, S. D. "Slime and Deposit Control in Alkaline Papermaking Systems." *TAPPI Papermakers Conference Proceedings*. 55-61. Atlanta: TAPPI PRESS, 1983.

32. Beach, E. M. 1984. *Tappi Journal*, 67(14):92.

33. McNamee, J. P. "Press Roll Considerations in Alkaline Papermaking." *TAPPI Alkaline Papermaking Seminar Notes*. 73. Atlanta: TAPPI PRESS, 1983.

3

Sizing With Alkenyl Succinic Anhydride

C. E. Farley and R. B. Wasser

Introduction

In the late 1970s and early 1980s, a strong trend toward alkaline papermaking developed. This was largely driven by the availability of low cost, high brightness calcium carbonate. The increased strength of an alkaline sheet permitted higher filler levels, with carbonate replacing clay, and titanium dioxide requirements reduced or eliminated. An alkaline conversion offered potential raw materials savings ranging from \$20 to \$50 per ton of paper (1).

With an alkaline sheet, an alternative to the conventional rosin-alum sizing system is required. This alternative involves a choice between alkyl ketene dimer (AKD) and alkenyl succinic anhydride (ASA) synthetic sizes. While each size has its advantages and disadvantages, ASA is preferred for many grades because of its on-machine sizing and size press holdout. Sizing with ASA is typically 80% to 100% developed on the machine, and paper can immediately be converted and shipped.

In many early alkaline trials and attempted conversions, ASA performance and runnability were very poor. Press picking, deposits and sheet defects caused many mills to terminate alkaline trials. The "rush to alkaline" was greatly slowed.

The problems with ASA application in early trials were real but were caused by a lack of thorough knowledge of ASA chemistry and on-machine application. The current understanding of ASA chemistry and advances in starch, polymer and emulsification technology have made ASA application relatively routine. Today there are approximately 60 machines in the U.S. running with ASA size, producing primarily fine papers and gypsum board.

ASA chemistry

Chemical composition

ASA is composed of an unsaturated hydrocarbon chain containing a pendant succinic anhydride. It is usually made in a two step process starting with an alpha olefin. The olefin is first isomerized by randomly moving the double bond from the alpha position. This gives an ASA which is a liquid at room temperature. In the second step the isomerate is reacted with an excess of maleic anhydride (the ene reaction) to give the final ASA structure as shown in Figure 3.1. Unreacted maleic anhydride and isomerate are stripped to give the final product.

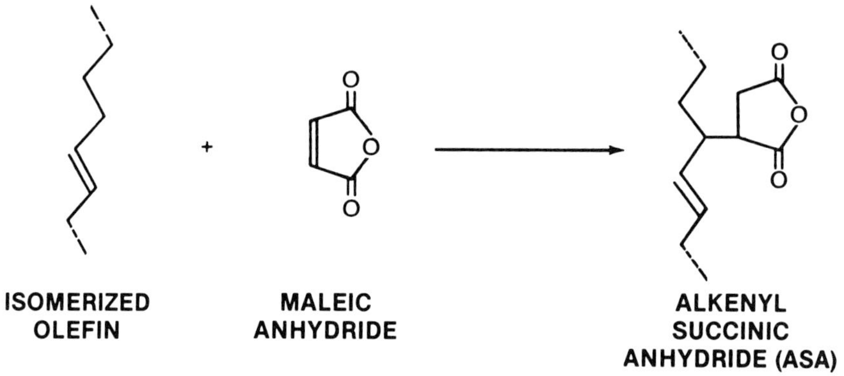

ISOMERIZED OLEFIN   **MALEIC ANHYDRIDE**   **ALKENYL SUCCINIC ANHYDRIDE (ASA)**

Figure 3.1 The preparation of ASA by the ene reaction between an isomerized olefin and maleic anhydride.

The starting alpha olefin is usually in the C-16 to C-20 range and may be either linear or branched. Sizing efficiency can be affected by the type of starting olefin used. In general, a longer chain length or a more linear chain gives a more efficient size.

Physical properties

ASA is an oily liquid with a light pale yellow color. Some typical physical properties are given in Table 3.1. Impurities are usually low and consistent mostly with residual olefin or maleic anhydride. Some suppliers add small amounts of an ASA soluble emulsifier which eliminates the need to add it separately when the ASA is emulsified at the mill. The shelf stability of ASA is excellent if protected from contacting water.

Table 3.1 Typical ASA properties.

Appearance	:	yellow to amber liquid
Density	:	8.0 pounds per gallon
Viscosity	:	150 to 160 cps at 23°C
Melting point	:	below 9°C

Chemical properties

ASA undergoes the usual reactions of anhydrides. Those of importance for sizing are esterification with the hydroxyls of cellulose and hydrolysis with water as shown in Figures 3.2 and 3.3. Reaction with cellulose is the generally accepted sizing mechanism (2,3). Hydrolysis is not desirable since hydrolyzed ASA is not an effective size. However, since the ASA is applied as an aqueous emulsion at the wet end of the paper machine, some hydrolysis is inevitable. Both reactions are relatively rapid which means that ASA emulsions must be used within a short period of time after preparation to prevent excessive hydrolysis. On the other hand, the high cellulose reactivity means that sizing develops early in the dryers and essentially full sizing is obtained at the reel or even before the size press.

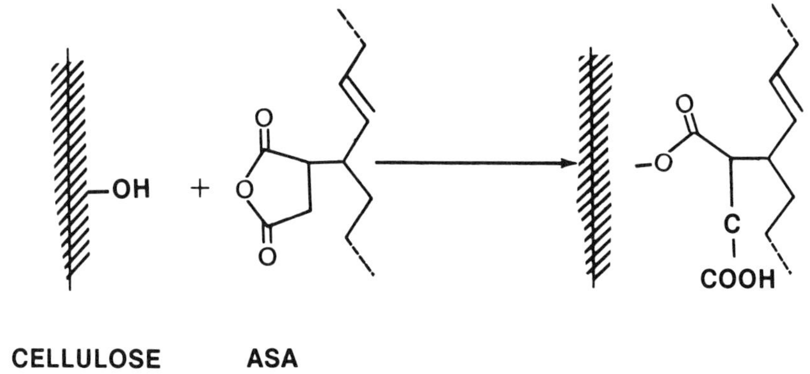

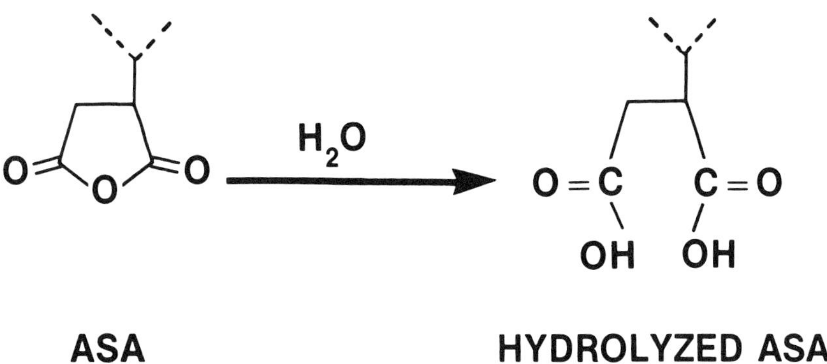

Figure 3.2 The esterification reactions of ASA with cellulose.

Figure 3.3 The hydrolysis of ASA.

The rate of hydrolysis is affected by the temperature and pH of the emulsion as shown in Figures 3.4 and 3.5. Hydrolysis can be minimized by keeping the temperature and pH during and after emulsification low and the time between preparation and use short. Hydrolysis can also take place after the addition of the size emulsion to the stock, and in fact, since both the stock temperature and pH are likely to be high, the rate of hydrolysis will be significantly increased. For this reason a point of addition should be chosen which keeps contact with the stock to a minimum. For the same reason, it is important to keep the retention of the size and the fines, onto which much of the size is adsorbed, high in order to prevent recirculation of size that has hydrolyzed during residence in the whitewater system.

ASA emulsion

Being water insoluble and of limited stability in water, ASA must be emulsified in the mill. Two methods of emulsification are available: low shear and high shear.

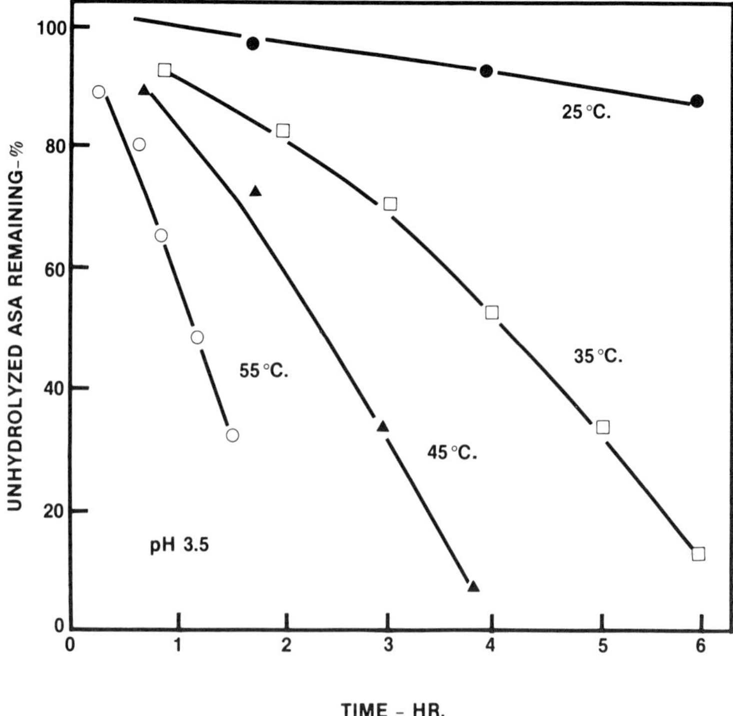

Figure 3.4 The effect of temperature on the hydrolysis rate of ASA emulsions at an initial pH of 3.5 (4).

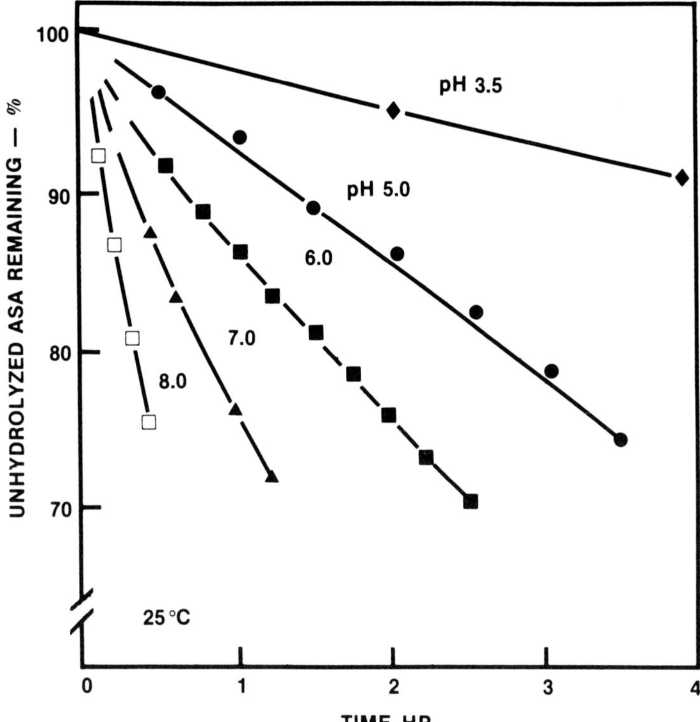

Figure 3.5 The effect of pH on the hydrolysis rate of ASA emulsions at 25°C (4).

Low shear emulsification involves passing ASA, starch, and a surfactant through a series of venturis (5). The surfactant is typically used at a level of 5% on ASA. This can result in foam problems and poor sizing efficiency due to the surfactant building up in the system and/or being retained in the sheet. Equipment wear due to abrasives can also be a problem.

High shear emulsification is most commonly employed. This involves passing ASA and a protective colloid, starch, or synthetic polymer, through a high shear turbine pump. All materials must be free of abrasives to avoid damaging the pump. A commercial high shear unit with flow sensors, controls, etc., is shown in Figure 3.6.

Most current ASA users emulsify the ASA in cationic starch using high shear equipment. Best results are obtained with a cationic potato starch. Potato starch gives an emulsion superior to corn starch in mechanical stability and sizing efficiency. Optimum ratio is 3:1 starch solids to ASA size. Higher ratios of starch to size give even better sizing efficiency, but press sticking (not picking) can result.

Synthetic polymers are available as alternatives to cationic starch for ASA emulsification (6). These give excellent and stable emulsions, but may have

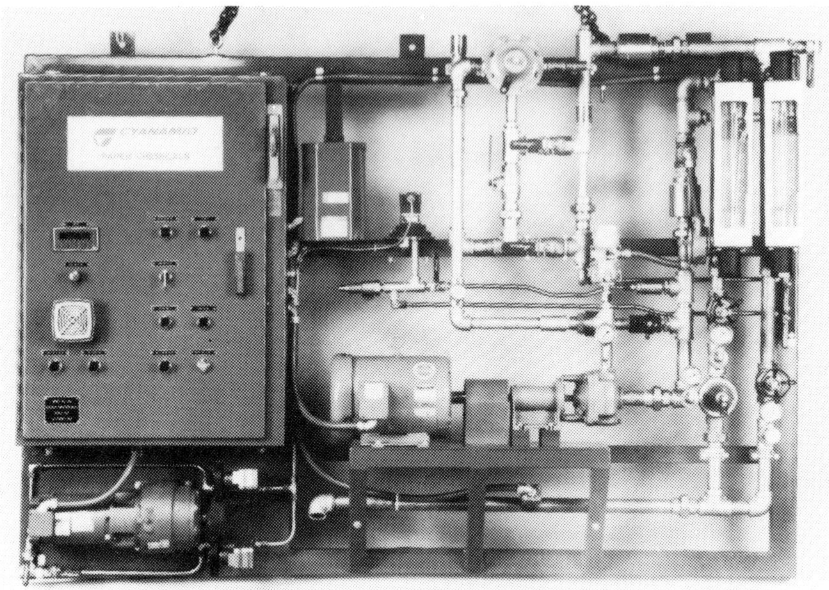

Figure 3.6. High shear ASA emulsification unit.

poorer sizing efficiency. Mills emulsifying with synthetic polymers may add cationic starch to the wet end to enhance sizing. In such cases, the polymer is an "add-on," as starch alone can be used for emulsification. The use of polymer instead of starch eliminates the incorporation of a nutrient into the sheet. This can be important in grades such as soap wrap and food board.

The optimum particle size for an ASA emulsion is about 0.5 to 3.0 microns. Technically, the smaller the particle size the better. Achieving this, however, may be impractical due to emulsion temperature and equipment limitations. Particles above about 3 microns tend to be unstable, and emulsions may cluster or break. This can cause deposit and/or runnability problems. Currently available equipment is capable of providing emulsion for good sizing efficiency and machine runnability (7).

ASA emulsions tend to be less stable when hard water is used. This is due to the presence of both calcium ion and alkalinity. Stability of the emulsion is enhanced when either cationic potato starch or synthetic polymer is used for emulsification. In hard water areas, storage of the ASA emulsion is not recommended, and direct feed to the paper machine is preferred.

ASA hydrolysis is the primary cause of any runnability problems. Therefore, it is imperative to minimize hydrolysis. This is most easily done by feeding ASA emulsion directly to the paper machine without storage. Where emulsion is to be stored, pH and temperature should be controlled. Emulsion pH should be reduced to about 3 to 4. This is most easily done by acidifying the starch or polymer, but addition of acid or alum to the emulsion has the same beneficial effect. Where starch is used for emulsification it should be as close to ordinary room temperature as possible. Some mills have installed heat exchangers to cool the starch to about 27°C. Again, problems with ASA hydrolysis due to emulsion pH, temperature and storage time are avoided by feeding the ASA emulsion directly to the paper machine.

A "good" ASA emulsion is mechanically stable against water hardness, shear, and extreme temperature and pH changes. Poor emulsion stability can result in deposits and poor machine runnability. The most stable ASA emulsions are those prepared in potato starch or synthetic polymers. Optimum ratios are about 3 to 1 starch solids: ASA and 0.4 to 1 polymer solids: ASA. These emulsions are stable even after extreme hydrolysis of ASA has occurred, a factor which enhances machine runnability and minimizes deposits. Emulsions made in corn starch are less stable and are further destabilized by ASA hydrolysis.

ASA application

ASA emulsion can be metered to the paper machine with any type of positive displacement pump. Addition to thin stock is preferred at any point where good immediate mixing is available. Suitable addition points are to the cleaner accepts or the screen inlet. Addition before the cleaners is not recommended as some ASA will be included with the rejects. This invites deposit problems due to hydrolysis of recirculating ASA.

The use of alum is strongly recommended with ASA size. Alum will enhance sizing efficiency, thereby reducing the amount of ASA required to achieve sizing tests (8). It is suspected that a reaction of ASA with aluminum hydroxide contributes to sizing. Alum also reduces the tackiness of hydrolyzed ASA, minimizing deposit problems (9). Recommended alum dosage is about 0.5% on paper production, or 5 kg per metric ton. Addition should be to thick stock in order to allow time for pH stabilization.

Good first pass retention is imperative in alkaline papermaking, particularly with ASA size (10). Poor retention results in ASA recirculating in the white water, hydrolysis and potential deposit problems. With state-of-the-art retention aids, high first pass retention is achieved (80% to 90%) and machine runnability is excellent.

Addition levels of ASA size will vary according to sizing specifications and filler levels. Typical doses for fine paper with an 18% calcium carbonate content, are shown below:

ASA	-	1 to 1.5 kg/metric ton.
Starch	-	3 to 4.5 kg/metric ton
Alum	-	4 to 5 kg/metric ton
Retention aid	-	as required

An alkaline sheet will behave quite differently from an acid sheet in the forming stage. Initial drainage will be rapid due to the tendency of calcium carbonate to shed water more easily than clay. This can have a negative effect on sheet formation, particularly in lightweight grades. Many alkaline mills compensate for this by removing up to one-third of the foils to retard drainage.

An alkaline sheet will have lower wet web strength than an acid sheet, due primarily to the higher filler levels used. In addition, fiber tends to hold more water under alkaline conditions. Reduced wet web strength has presented some problems only on lightweight grades made on machines without pickup felts.

Whether running acid, neutral or alkaline, all paper machines will require some downtime for clean-ups or boilouts. The boilout frequency will vary for each particular machine. Alkaline boilouts have been found to be the most effective for machines running ASA size. Adding the boilout chemicals to regular stock is recommended in order to take advantage of the "abrasiveness" of the stock and to carry the boilout chemicals onto the forming table. Boilouts should be thorough, as their purpose is to remove deposits, not to soften them. Optimizing the boilout frequency and duration is an evolutionary process and will vary with each machine. As a rough guide, most machines running ASA are boiled out at frequencies ranging from two to eight weeks.

Potential problems

Until very recently, many mills rushed into ASA trials without adequate planning or understanding of ASA chemistry. Both the mills and the suppliers were at fault. The results were the disasters we have all heard about - deposits, press picking, holes, bad paper, etc. Let us now look at why these problems occurred, and at how recent advances in ASA application technology has eliminated them.

Virtually all problems with ASA size follow from poor emulsion stability or from excessive ASA hydrolysis.

In the mid 1970s, one of the authors had the painful experience of conducting several trials where ASA was emulsified in water and surfactant only. No starch, no polymer, no protective colloid and little stability to the hardness of the alkaline environment. Poor emulsion stability led to almost immediate press picking and feed line deposits. Use of rosin sized broke only compounded the machine problems. In the worst case, where ASA was emulsified in very hard water, emulsion breakage occurred and resulted in "plating" of the machine wire with ASA, in effect sizing portions of the wire. This caused "craters" of wet stock to run down the table without draining.

Those were the early days of ASA technology. Recent research had led to the development of specialty potato starches and synthetic polymers for ASA emulsification (6,11). These give excellent stability of ASA emulsions in hard water and alkaline environments, virtually eliminating press picking and machine deposits. Optimum ratios for good emulsion stability are about 3 to 1 potato starch to ASA and about 0.4 to 1 polymer solids to ASA.

Since ASA hydrolysis is a major concern, emulsion storage time should be minimized or eliminated. With state-of-the-art emulsification equipment, emulsion storage is justified only to provide time for equipment maintenance or repair. When ASA emulsion is stored, the pH should be in the 3 to 4 range. Sulfamic acid, mineral acids or alum can be used.

Some ASA hydrolysis is inevitable, and addition of alum is recommended to eliminate any press picking or deposit forming tendencies of hydrolyzed ASA. Addition of about 0.5% alum, or 5 kg/metric ton, is effective. Addition should be to thick stock (machine chest) to minimize any machine pH control or foam problems.

Optimizing first pass retention is mandatory when using any synthetic size. Unretained and recirculating size will inevitably hydrolyze and invite runnability problems. Careful selection of a retention aid is therefore an important part of planning the alkaline conversion. Anionic retention aids are synergistic with the cationic potato starches and polymers used to emulsify ASA. However, "picking the winner" still involves screening of retention aid candidates in the laboratory, e.g., with a BRITT jar to simulate paper machine shear.

The most important part in planning an alkaline conversion is *Planning.* Effective planning has one prerequisite, and that is commitment. Whether the motive be job preservation, profit sharing or pride, the commitment must be there. Everyone, from machine crews on up, must be willing to learn to make paper in a different way. Supplier seminars should be held in order to make mill personnel as educated as possible. What will change and how do you react? These questions are effectively answered only in the planning stages of the alkaline conversion.

Case histories

Alkaline papermaking involves many changes other than simply raising the machine pH. Consider the following results of a conversion to alkaline:

- refining efficiency is changed
- drainage is changed
- drying is changed
- fillers and levels are changed
- retention is changed
- sheet bulk and porosity are changed.

With the aid of modern computers, the experienced papermaker can react to these changes, especially if he expects them. Beyond pure papermaking, changes may also be required in overall operation, biocides, defoamers, dispersants, felt washes, surface sizes, etc. The key to anticipating these changes and to employing effective reactions is a *total commitment to the alkaline conversion.* Thorough contingency planning will minimize most of the problems, and a successful conversion to alkaline papermaking can be expected.

The differences between poor planning and good planning are illustrated in the following examples.

Poor planning: I

A board mill decided to "look at" alkaline papermaking (no real commitment). The incentive was the reduced basis weight and reduced raw materials costs being enjoyed by a competitor. The trial plan was simple, to gradually raise the machine pH to about 6 and then substitute ASA for rosin size. When the pH rose above 7, about 5% calcium carbonate was to be added. This may seem to be a fine approach, but consider what happened. Foam, both on the machine and in the clarifier (mill is 100% closed) became unmanageable. The trial was aborted before pH 6.5 could be reached.

The effect of unretained rosin in the white water and clarifier was never considered. Soap size, as we know, is an excellent foaming agent particularly at high pH. This was not considered in advance, and the result was predictable. Whether deposit problems from calcium rosinate would have been a problem was never determined, as the machine pH was never raised sufficiently to permit carbonate addition. To further emphasize the lack of planning, the mill ran two more identical trials with the same results. Dumping the clarifier would have been less expensive than the lost machine time, but the mill did not have the commitment to do this.

Poor planning: II

A fine paper mill was ordered by corporate to convert to alkaline papermaking. About 6 months notice was given. Planning for this conversion consisted essentially of the following:

a) Selecting an ASA supplier,
b) Installing ASA storage, emulsification and metering equipment, and
c) Installing cationic starch cooling equipment - the mill already used cationic starch.

On start-up, ASA emulsion was added to the top of the stuff box. Mixing with stock at this point was very poor (a puddle of emulsion collected at a stuff box corner), and caused poor ASA sizing efficiency. This in turn increased the ASA requirement, so that an excessive amount of ASA was being added.

Press picking, machine deposits and sheet defects were constant problems, and resulted in the mill calling in other synthetic size suppliers.

Sizing variations were extreme, with HST values being nil for several reels and then very high. Tests were commonly recorded as either zero or "300 seconds," the latter being when high tests were terminated. The testing ink was found to be unsuitable with ASA. When any sizing at all existed, the ink would not penetrate the ASA-sized paper.

These and other problems were eventually solved, but 6 or 7 months passed before the mill reached normal production rates.

Good planning

A fine paper mill made the commitment to convert to alkaline papermaking with ASA.

Several ASA suppliers were invited to discuss their products, experience and recommendations to key production and technical personnel. At least one was invited back several times to review their approaches and to answer questions. Based on these and internal meetings, an ASA supplier was chosen - a single supplier for ASA, cationic starch and retention aid.

Installation of permanent-type equipment was begun 3 months before the planned conversion. Everything was wired to the main control room. An on-site precipitated calcium carbonate plant was being built.

Britt Jar tests to select the preferred retention aid were begun, with mill and supplier personnel working together. Testing was repeated many times in order to be certain the selected polymer was not sensitive to ordinary furnish variations or grade mix.

Addition points for ASA emulsion were installed. All equipment was tested under the expected operating conditions.

Every detail was discussed in continuing meetings between mill and supplier. Meetings were also held with the machine crews and Quality Control people to discuss machine operating parameters.

All of the planning and preparation paid off as the second reel of alkaline paper met specifications. Some problems with wet web strength and sheet porosity in lightweight grades were later encountered, but these were solved. The mill is now realizing essentially all of the originally estimated cost savings with alkaline papermaking.

Conclusion

A conversion to alkaline papermaking with an ASA size offers the papermaker an opportunity for enormous savings in raw material and production costs. However, it also risks some serious potential problems.

Of course, one would prefer to avoid all potential problems in the first place, but this is rarely the case. However, understanding ASA chemistry and alkaline papermaking to the fullest extent possible beforehand will prevent the potentially fatal problems. It will also allow for what problems that do occur to be solved or reduced to only nuisance levels in the shortest possible time.

In making the smoothest possible conversion to alkaline papermaking, there are two essential elements. These are planning and commitment, and apply to both the papermaker and the supplier. All aspects of the initial trial work should be planned in detail, and nothing should be assumed. Changes in retention aid, biocide, dyes, defoamer, etc., may be required and should be anticipated.

Production and quality control people should become involved as early as possible. All should understand the whats and whys of the chemical and mechanical changes to be made. A mill-wide commitment must be made to make the conversion a success.

References

1. Maher, J. E. 1983. *Pulp and Paper.* 57 (6):118.
2. McCarthy, W. R. and Stratton, R. A. 1987. *Tappi Journal.* 70 (12):117.
3. Dumas, D. H. 1981. "An Overview of Cellulose Reactive Sizing." *Tappi.* 64(1):43.
4. Wasser, R. B. *Alkaline Papermaking Seminar Notes.* 17. Atlanta: TAPPI PRESS, 1985.
5. Mazzarella, E. D., et al. U.S. Pat. 4,040,900 (1977).
6. Rende, D.S., et al. U.S. Pat. 4,657, 946 (1987).
7. Farley, C. E. *Alkaline Papermaking Seminar Notes.* Atlanta: TAPPI PRESS, 1985.
8. Strazdins, E. *Alkaline Papermaking Seminar Notes.* Atlanta: TAPPI PRESS, 1985.
9. Scalfarotto, R. E. 1985. *Pulp and Paper,* 59, p. 126.
10. Maher, J. E. *Alkaline Papermaking Seminar Notes.* 89. Atlanta: TAPPI PRESS, 1985.
11. Farley, C. E. *Sizing Short Course Notes.* Atlanta: TAPPI PRESS, 1987.

4
Surface Sizing

Dale R. Dill and Kenneth A. Pollart

Surface sizing

Surface sizing is a generic term which simply means the application of materials, usually starch, at the size press. Consequently, when the industry talks about surface sizing they usually are not referring to water resistance, but rather to the more general topic of applications to the surface of paper. This discussion will deal with surface sizing in general, and then more specifically the addition of additives to the size press formulation to impart resistance to liquids including water. Surface sizing applications typically involve formulations with 30% or less solids generally composed of starch or proteins along with fillers and/or various additives to achieve the particular properties desired. When solids are higher than 30%, although this is not a clear-cut dividing point, one would more likely refer to the application as a coating.

Pick-up

The percent pick-up by the surface application refers to the quantity of size press formulation applied from this surface application compared to the total weight of the paper entering the size press. For most internally sized grades one normally experiences a 30% to 50% wet pick-up at the size press. By that, we mean the application of 300 to 500 lbs. of size press formulation per metric ton of paper. Levels of pick-up as high as 100% or 1,000 lbs. of formulation per metric ton of paper might be achieved with a totally unsized base stock on a very slow machine.

Porosity

Surface sizing is used for the development or changing of the porosity of the base stock. Historically the most common test for paper porosity has been the Gurley porosity tester. This instrument measures the time required for 100 cc of air to flow through a sheet of paper. As the permeability of the paper is decreased, the Gurley time units increase as the air is restricted from flowing through the sheet. If the porosity of a sheet is increased, technically that means that the ability of air to flow through the sheet has increased and the Gurley porosity reading decreased. If, on the other hand, the porosity is decreased by making the sheet more closed by refining or application of resins, then the Gurley porosity reading is increased. Thus, it is always advisable for those discussing changes in porosity to clarify what is intended and the terms each person is using to define the change in porosity.

Mechanisms of size press pick-up

An understanding of size press applications is not simple. One must understand nip applications, the hydrodynamics of the size press and their interaction with the base stock. Toward the goal of gaining a better understanding to aid the papermaker, a paper was presented at the 1973 TAPPI Coating Conference which was later published in *Tappi Journal* in January, 1974. A portion of that work is present here in the belief that it does contribute to a broader understanding of size press applications. It explains some of the phenomena that we see in daily operations of the size press, and it can help predict changes on the paper machine and in the base stock after size press applications. This work was done on two pilot paper machines - one at Western Michigan University and the other at the Mead Corporation in Chillicothe, Ohio, where size press pick-ups at speeds varying from 30 meter per minute to 900 meter per minute were measured. The influence of several variables on pick-up control with size press formulations whose viscosities ranged from 6 centipoise to nearly 100 centipoise were studied. The primary variables in getting this viscosity range was starch solids and starch molecular weight. Three base stocks were used for the principal study - all were 50/50 hardwood/ softwood kraft at 73 gm per square meter basis weight:

    Unsized and very open.

    Unsized, but a tighter base sheet with a Gurley porosity of 30 seconds per hundred cubic centimeters.

    A sized sheet that was very opened like the first.

These three preformed based sheets were run through the size press with the various formulations at varying speeds. Pick-up was determined by taking tear-outs prior to and immediately after the size press.

On extended runs in the mill, with relatively constant conditions of basis weight, speed, etc., a reasonable approximation of pick-up can be determined by measuring the draw-down from a size press delivery tank as compared to the production from that machine in tons at the reel. For this method to be reasonably accurate, it needs to be done over a long enough time period that the change in volume of the delivery tank can be accurately measured.

Internal size is a prime factor in control of size press pick-up at slow speeds and/or with low viscosity solutions. At higher speeds with more viscous solutions the pick-up of a sized sheet approaches that of a waterleaf sheet. In all cases with a sized base sheet an increase in machine speed increases pick-up.

Figure 4.1 demonstrates how little sizing is needed in the paper web as it enters the size press to control pick-up. The level of sizing needed is too low to be measured by the standard ink penetration tests. This is shown in Figure 4.2 where arbitrary sizing levels are related to pick-up. By definition, at a sizing level of zero there is known to be no sizing agent added; at a level of 1, sizing agents are present but not detectable by standard ink penetration sizing tests; at a level of 2 a hint of sizing exists by these tests; and at a level of 3 a definite but low level of sizing is present. Size press pick-up is altered by sizing level 1 and completely controlled by level 2. This indicates that the size press is a very sensitive test for low levels of sizing.

While recognized in the laboratory that a more dense waterleaf sheet required longer time for saturation, the magnitude of the response was not known. Many grades today use a waterleaf sheet of high Gurley porosity. Most of these grades require more surface oriented filming applications. This study shows that an increase in Gurley porosity aids film formation, not only by reducing the void area between fibers making film formation easier, but also by better holdout of the size press formulation. At high speeds, pick-up is influenced more by the sheet porosity than by internal sizing.

It is also recognized that speed affects pick-up, but the magnitude of its

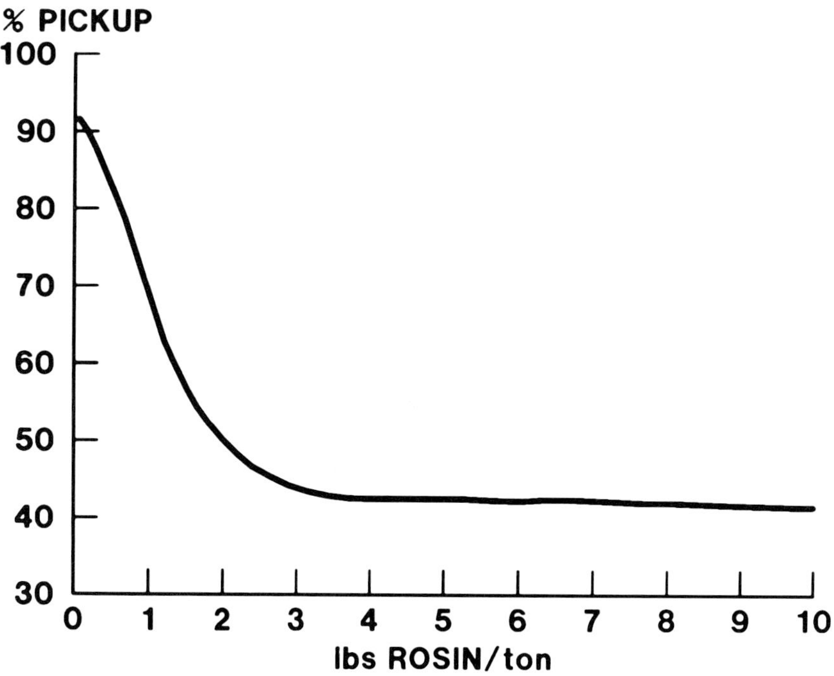

Figure 4.1 Percent pick-up vs. level of internal rosin size.

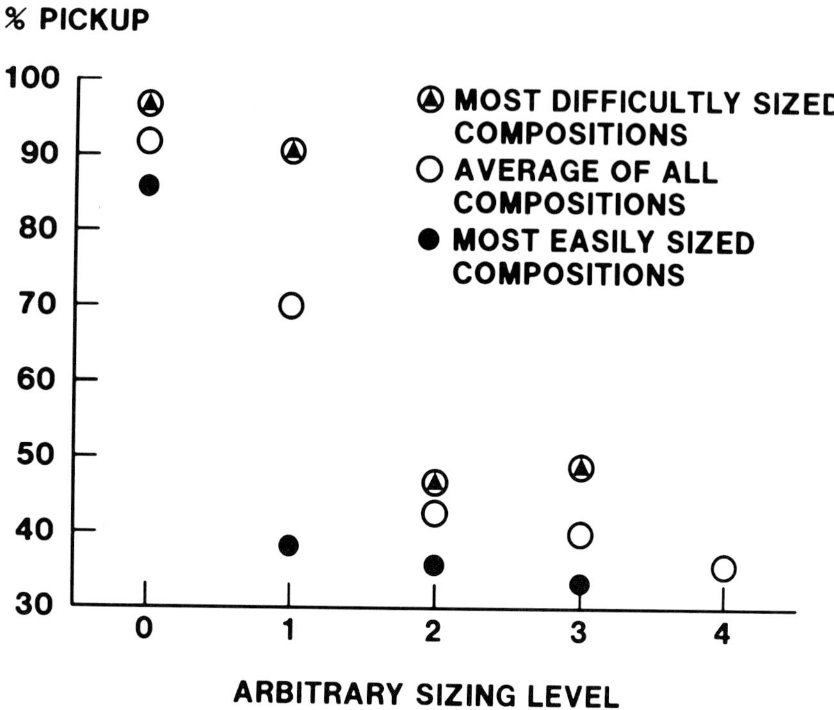

Figure 4.2 Percent pick-up vs. sizing level of basestock.

effect was again underestimated. The effect of speed is so interrelated with viscosity and base sheet properties that one can only say that speed significantly affects size press pick-up (Figures 4.3 and 4.4).

Starch viscosity, as affected by molecular weight of the starch molecules (or degree of conversion) and solids content, contributes enormously to variations in pick-up. At a low viscosity a given base sheet may decrease in pick-up with speed, while at a higher viscosity, pick-up can increase with speed. Viscosity plays a key but varied role in all the mechanisms involved.

Other factors do affect size press pick-up. However, it would appear that, except in extreme cases, factors such as moisture into the size press may play lesser roles. This is not to be confused with their effect on other factors such as uniformity of pick-up.

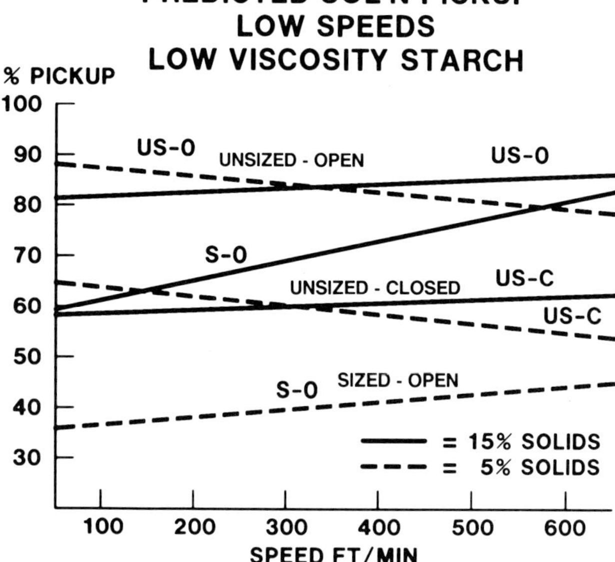

Figure 4.3 Predicted sol'n pick-up low speeds low viscosity starch.

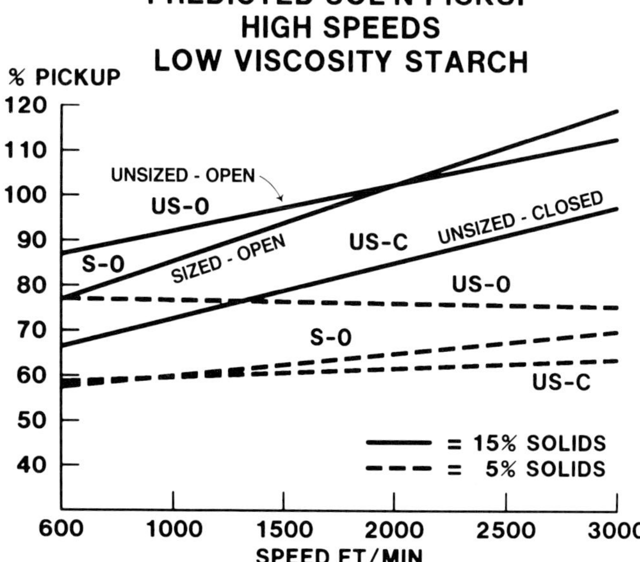

Figure 4.4 Predicted sol'n pick-up high speeds low viscosity starch.

Theoretical considerations

Wet pick-up at the size press would appear to be controlled by four principal mechanisms (equations included) depicted in Figure 4.5. These are:

1. Capillary penetration in the pond,

$$= k \, \frac{r \gamma \cos^{\theta\frac{1}{2}} t}{n} = k' \, \frac{r \gamma \cos^{\theta\frac{1}{2}}}{ns}$$

2. Pressure penetration up to the nip midpoint,

$$K = \left(\frac{P}{n}\right) t = K' \left(\frac{P'}{ns}\right)$$

3. Liquid metered through the nip by surface roughness

$$= k \times \text{roughness}$$

4. Free film on the surface due to hydrodynamic metering and film splitting.

$$= k \left(\frac{ns}{P_o}\right)$$

Where:

r	=	radius
γ	=	surface tension
$\cos\theta$	=	contact angle (90°)
n	=	viscosity
k	=	constant
t	=	time
s	=	machine speed
P_o	=	pressure-gauge
p	=	applied pressure at any given point in the nip

In conclusion, the level of internal sizing is the greatest factor in size press pick-up control. However, control of size press pick-up can be achieved with very low levels of internal sizing which are undetectable by ordinary sizing tests.

The viscosity of the size press formulation, as controlled by concentration and molecular weight of the starch, also has a strong influence on size press pick-up. The magnitude of this effect is very dependent upon the base stock and can, in fact, cause an increase or decrease in pick-up.

The porosity of the base sheet, while not always an operating control, does have a large effect on pick-up.

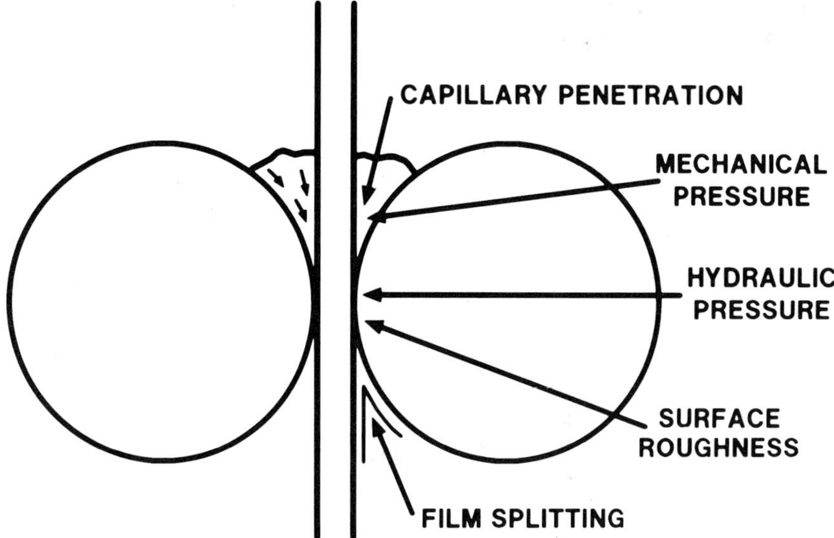

Figure 4.5 Factors involved in size press penetration.

Machine speed influences pick-up through multiple avenues. Like viscosity, the degree and the direction of its influence depends on the base stock characteristics. Increased machine speed can increase or decrease size press pick-up, depending on the speed and the base stock.

The pick-up on an internally sized base sheet is influenced more by operating variables, (e.g. speed and viscosity) than is a waterleaf sheet.

While at low speeds an unsized sheet picks up more volume of size press formulation than a sized sheet; at high speeds (e.g., 600 meter per minute), there may be a little significant difference.

This has been a rather cursory and fragmented presentation of these studies, but it should serve to give some improved understanding of the roles played by internal sizing, speed, porosity, and solution viscosity in control of size press pick-up. A more detailed and complete explanation and presentation of data can be found in *Tappi*, 57 No. 1, 97-100 (1974).

Functional property response

Water resistance

The most common performance sought from size press additives is water resistance or sizing. The level and type of water resistance desired will vary with the grade of paper or board as well as with the type of printing (letter press printing would require less than offset). It should also be kept in mind that water penetration resistance and water pickup resistance will not necessarily be obtained with the same materials. For example, the application of starch alone at the size press will generally provide some improvement in water resistance as measured by an ink penetration test; however, it will hurt water resistance as measured by a water pickup test (such as Cobb). Low viscosity starches can even cause an internally sized sheet to fail a pen and ink feathering test. Thus, it is important to utilize a water resistance test that will reflect the type of end use the sheet is designed for.

If a significant increase in water resistance is desired from the surface sizing application, then it is recommended that one of the many additives designed for this use be utilized along with the starch. It is generally recognized that additives incorporating long alkyl chains or even wax itself will provide very high levels of water resistance - even a non- wettable surface if desired. However, if a more controlled level of water resistance is desired or if a surface is needed that can be readily and uniformly wet by a fountain solution or an aqueous coating, then it may be more desirable to utilize one of the additives that have been designed to provide a compromise between maximum water resistance and other desirable sheet properties.

The most efficient method of achieving water resistance is to use a combination of a low level of internal size and a controlled level of a surface sizing agent. This approach provides the papermaker a great deal of wet-end freedom. The level of alum in the system can be 0 to 20 to 25 kg/metric ton. Any internal size can be used in conjunction with a surface sizing agent. The pH of the wet end can be from 4 to 9, depending upon the internal size chosen. Any filler desired can be used in the wet-end system, but just as with wet-end sizing, the higher the level of filler in a sheet, generally the higher the amount of additives that will be necessary to achieve a given level of water resistance.

Wet rub and wet pick resistance

Wet rub or wet abrasion resistance is generally obtained by the incorporation of a crosslinking type of additive. These materials typically do not provide any increase in water penetration resistance and little or no response in wet pick resistance. Wet pick resistance is usually obtained by utilizing one of the conventional sizing type additives discussed previously. Wet rub is normally not critical on uncoated papers, but wet pick can be important on the uncoated offset grades. For many printed grades, including offset, suitable printability can be achieved by an appropriate level of either wet rub or water penetration resistance. In those cases, the sheet prints satisfactorily whether it has been wetted and retained its strength or whether it has simply resisted wetting and thus had sufficient strength.

Internal strength

Improvements in internal bond strength, as measured by Z-direction strength tests, is one of the primary contributions of size press applied starch. The level of internal strength improvement depends both on the degree of penetration of the starch and the type of starch used. The low molecular weight, brittle starches are generally weaker than the higher molecular weight or derivatized starches. The strength response of the starch can be improved by the addition of one of the strong hydrogen bonding polar types of polymers that are available from a number of suppliers. Polymers of this type normally contain hydroxyl, amine, carboxyl and/or amide functionality. Much of this potential contribution to internal strength is lost if the formulation is applied with one of the roll or blade types of coaters rather than a conventional size press due to the more surface oriented placement resulting from these applicators.

Surface strength

Size press applied starch can be utilized to control linting, dusting and picking. The proper selection of the starch is much more critical for surface strength response than it is for internal strength. The over-converted, low molecular weight starches can be so brittle that they actually contribute to linting and dusting rather than prevent it. By the proper selection of a good film forming starch, often in combination with one of the polymeric additives discussed above (internal strength), many mills have obtained major improvements in these properties. For example, this approach has been utilized for years to control vessel segment picking and more recently has become particularly important in the highly filled, alkaline-sized sheets and with sheets containing high levels of mechanical fibers.

Porosity

The open pore structure of the sheets can be partially closed, thus porosity reduced, by the filming action of the starch or other materials applied at the size press. The degree of response is very dependent upon the nature of the base sheet, the filming characteristics of the material being applied, the pounds of material applied, and the method of application.

If the base sheet is very open, Gurley porosity of 10 sec. or less, then it is very difficult to obtain more than a few seconds improved response from the size press formulation. If the initial Gurley porosity is 15 to 20 sec., then it is possible to achieve modest improvements with starch alone and up to 80 to 100 or more sec. with starch plus a film forming additive. Likewise, if the initial Gurley porosity is in the hundreds then increases to the thousands can be readily obtained.

Since this porosity response is dependent upon obtaining a continuous film across a maximum number of the pores in the sheet and maintaining the integrity of those films through the calendering and other mechanical handling of the sheet, it is important to utilize a starch with good film forming characteristics. This means using starches of higher molecular weight and/ or derivatized starches. The response of any of these starches can be substantially enhanced by the addition of one of the polymeric film-forming additives that are available. In most cases addition levels of 10% to 25%, based on starch, are required to obtain significant changes in response. In some cases, the economics of the grade justify the use of the synthetic polymers by themselves.

Roll or blade types of coaters, which tend to give more of a surface orientated placement of the applied material, are often preferred over conventional size presses where maximum filming and porosity response is desired.

Non-aqueous holdout

The holdout of non-aqueous materials can be achieved either by treating the sheet with an oleophobic agent, such as the fluorocarbons which are discussed in another chapter of this book, or by utilizing solvent insoluble, film forming materials such as starch and/or synthetic polymers as discussed in the previous section on porosity. One of the advantages to the film-forming approach to non-aqueous holdout is that the desired level of solvent holdout can be obtained while still maintaining a sheet surface that is wettable by the solvent. This is especially important in solvent coating applications where a uniformly distributed and firmly bonded coating is desired. Likewise holdout achieved by film forming is less subject to failure from pressure.

Ink receptivity and show-through

The receptivity of oil-based inks is primarily controlled by the porosity of the sheet structure. However, the degree of ink penetration into the sheet, as well as the surface wicking of the ink, can be readily influenced by the size press formulation. Starch by itself will often provide adequate ink penetration control. A major improvement in ink gloss and brightness along with improved sharpness of the image characters can be obtained by the addition of relatively small amounts of one of the film-forming polymeric additives that are available. By controlling the ink penetration, this approach also tends to minimize ink show- through.

Opacity

In general, the size press is a detractor from opacity since the material applied tends to fill voids in the sheet and make it more transparent. This negative effect can be minimized by controlling the pounds of material applied at the size press and by adding pigment to the formulation.

On those rare grades where a more transparent sheet is desired, then the size press formulation should utilize low viscosity materials so maximum pick-up and penetration can be obtained.

Problems encountered

Foam

Foam control is essential for satisfactory size press operation. The presence of uncontrolled foam in the size press formulation has a number of negative effects. It increases the apparent viscosity of the formulation (especially at the point of application), causes changes in pickup and placement (both because of viscosity and because of the reduced solids in the volume picked up by the sheet), affects the uniformity of the application during the run as the amount of foam changes with time, and affects the distribution of the applied material.

The presence or absence of foam is typically "determined" by observing the formulation at the point of application and in the supply tanks. A more reliable procedure, since fine entrained air is difficult to see, is to compare the density of the formulation returning from the application point and that of the foam-free formulation. This is satisfactorily accomplished by simply weighing approximately equal volumes of the two samples.

The first step towards foam control should always be the minimization of air entrainment due to cascading in recirculation systems, violent agitation, and air leakage in pumps. The second step is the judicious use of foam control agents.

There are two factors that must always be kept in mind when selecting a foam control agent. The first, of course, is does the agent control the foam of the specific formulation being used. The second, which is most often given very little attention, is does the foam control agent interfere with the performance of other critical additives or the final desired performance of the sheet. The most common interference occurs between the "surface active" components of the foam control agent and the aqueous resistance response of sizing additives. An improperly chosen defoamer can, in fact, destroy all or part of the water resistance gained by internal or surface sizes. However, with the variety of products available today, finding a satisfactory performing yet compatible system should not be a major problem.

Surface slipperiness

Excessive surface slipperiness is most often encountered as an unwanted side effect from an attempt to obtain maximum water resistance, especially with the long alkyl chain or waxy types of additives. The problem can best be avoided by carefully controlling the use level of the additive contributing to the problem or by selecting a sizing additive that does not have this side effect.

Important properties of size press formulation

The properties of the size press formulations affect the paper properties achieved from them. As demonstrated in the size press pick-up studies, the viscosity and rheology of the size press solution are critical to the operation of the size press. Consequently, as additives are put into a size press formulation, corrections must be made for viscosity and rheology. One must ensure that all ingredients in the size press formulation are compatible. Compatible not only with the property desired (for example one cannot add surfactants to the size press if water resistance is needed in the final sheet) but, also with the chemistries present (for example highly anionic and cationic materials or other mutually reactive species are generally incompatible). Certainly, there are other properties which are important, but this short list should suffice to indicate the types of considerations which must be given to size press formulations.

Combinations of size press additives

Probably the least perfected area of size press technology is the use of combinations of functional additives. Starch is typically applied to the surface of paper for those well- known reasons of laying surface fuzz and bonding the surface fibers well to the sheet. But in many cases this causes a problem by putting a very hydrophilic layer on the surface. A combination of starch and a sizing agent can solve that problem and lead to much better properties such as Cobb or pen & ink feather. Fluorochemicals are used to impart high levels of grease resistance, but in many cases fall short on the level of water resistance provided. A combination of a fluorochemical with one of the sizing agents can provide both grease resistance and water resistance. Many film formers such as polyvinyl alcohols and sodium alginates are used to increase the Gurley porosity of the paper web, but again provide little or no water resistance. If one uses a combination of a film former with a material that provides water resistance, an excellent balance of properties is achieved. This list of combinations could go on and on. The purpose is not to describe every such opportunity but rather to note that combinations of functional additives can be used to achieve a balance in properties not achievable with either additive alone.

Interaction with subsequent coatings

Key also is the interaction that size press applications have with the success of subsequent coatings. A properly selected size press formulation can greatly improve the results of a subsequent calender stack application. The size press application prepares a base, imparts some water resistance, making the calender stack application easier to be held near the surface, and impart its properties to a much higher degree. Size press sizing can also be advantageously utilized to impart coating holdout for subsequent high solids pigmented coatings. The papermaker has experienced the dewatering of aqueous high solids coatings (with inadequate water holding capacity) into the base sheet, causing a variety of runnability and/or final sheet deficiencies. One approach to minimizing this problem is to build a level of water holdout resistance into the basesheet with a size press application. Thus, dewatering into the sheet is slowed, solids of the coating are maintained, the binders stay more in the top coat rather than strikin into the base sheet, and the coating is stronger and more trouble free. In some cases, reformulation of the coating may still be necessary to obtain a totally satisfactory solution used for silicone release coatings, the proper preparation of a base sheet by size press application prevents or at least minimizes the strike-in of the more expensive silicone coating and not only improves performance but reduces overall cost.

Again, the goal is not to itemize such possibilities for size press/coatings interactions but rather to develop the thought which says, "When I am applying a coating, and I am experiencing problems such as striking, dewatering or rheology change with time, I should consider the size press as a way to correct the problem rather than changing the coating or the wet end design of the base sheet." Frequently, that thought can provide improved costs and quality with a minimum of change and effort.

Effect of higher speeds

There is an obvious trend toward increased machine speeds. The data discussed earlier quantifies the effects that increased machine speeds have on size press pick-up, especially in 300 to 900 meter/min. or higher range. There can be dramatic increase in the pick-up at the size press with these increasing machine speeds. For example, the size press pick-up likely will be increased by 20% or more as one goes from 600 to 900 meter/min. Consequently, to maintain the same solids pick-up one will have to reduce the solids of the size press formulation. These increased machine speeds require lower and lower viscosities to avoid nip rejection. To achieve the needed lower viscosities one would normally convert the starch to a greater extent. Doing so reduces the molecular weight and subsequently the positive influence that the starch will have on strength. One must be even more careful about the influence of size press additives on the viscosity of the size press solution. As the speed is increased, the pick-up is more and more controlled by film splitting which tends to level the materials closer to the surface of the paper. This is good if one is looking for surface properties or for film formations, but not if one is hoping to achieve improved 'Z' direction strength.

On lower speed machines, the use of internal size is the primary method used to control size press pick-up. As machine speeds increase, the difference in the pick-up between sized and unsized paper diminishes. Thus, there comes a point then where internal size is not effective in controlling size press pick-up and other means must be employed. In fact, in many such cases, internal size probably can be eliminated. These changes due to increased speed have caused a renewal of interest in alternate methods of applying size press formulations such as gate roll, Billblade (R) and short dwell coaters. This equipment does minimize or eliminate control problems such as increased size press pick-up and nip rejection. On the other hand, these application techniques do not allow deep penetration of the formulations. This can be a positive or a negative feature depending if surface filming or internal strength are the primary responses needed from the size press.

Effect of additives on drying

Do size press additives affect the drying after the size press addition? Most size press additives, at the low levels used, do not alter the filming characteristics of the starch. However, some size press additives at high levels of addition do increase the filming response. Only if the size press additive does cause significant filming which makes the release of water more difficult or if the size press additive significantly increases the viscosity (either directly or indirectly by foam generation), thus increasing the size press pick-up, will drying be adversely affected. Consequently, one can be relatively sure that most size press additives do not affect drying. Most viscosity increases can be reduced to a normal level by changing solids or using a lower viscosity starch, so increases in size press pick-up due to viscosity can be corrected.

If heavy filming is desired and obtained, some increase in drying demand can be expected. In all other cases it should be possible to take corrective steps that will avoid increased drying demands while achieving the desired property from the size press additive.

Size press additives available

It would be ideal to list the numerous products used at the size press, describe their chemistry and how they can best be used and who supplies them. To be fair that list would have to be complete and accurate. One soon realizes that would be an impossible task. There are a variety of products recommended for use at the size press encompassing a wide variety of chemistries. They are promoted to achieve the host of performance characteristics earlier described as well as others. Generally they can be divided into film formers and non-film formers which isn't the same as polymeric and monomeric, as many polymers do not form films especially at the levels used. There are excellent products to be used at the size press supported by excellent suppliers that understand their products, what they will do and how to use them. Those products and suppliers can be located through the standard industry references which list such suppliers.

Advantages and disadvantages

As would be expected, there are advantages and disadvantages to surface sizing. It would be our opinion that the advantages far out-weigh the disadvantages, but it serves a useful purpose to discuss both. Some of the many advantages to the utilization of size press applications include:

1. Size press applications are very efficient.
   a. The size press solids are 100% retained. Only the volatile components are lost during drying as compared to the wet end where a significant portion of the desirable additives frequently remain in the white water and subsequently are lost from the system.
   b. The material is applied to the surface. This provides the opportunity for controlled quantity and placement of the applied materials. For example, in those applications where only a surface response is needed, such as ilming for porosity or non-aqueous holdout, the formulation and application techniques can be selected so as to keep the majority of the applied material at the surface thus obtaining a maximum response at a minimum cost.
2. Size press applications minimize contamination of the environment and consequently give rise to low environmental concern. This is not only because they are 100% retained, but also because the quantities of size press formulation can be limited to the run tank if it is used as the point of addition for the additive. Thus at the end of a run only a minimal amount of material need to be discarded.
3. Size press chemicals can be simpler in chemistry and use because one need not worry about retention and other factors that are critical in wet-end applications. Consequently, a wider variety of materials are available for use at the size press.
4. The size press chemicals generally do not interfere with fiber to fiber bonding as do many of the wet-end chemicals since the bonds have already been formed when the size press formulations are applied.

Some of the limitations and disadvantages to the utilization of size press applications are:

1. On an internally sized base stock and even on many of the heavier weight waterleaf base stocks, complete penetration is not achieved. Therefore, the internal portion of the web in those situations is not altered by size press applications. For example, only the outer portions of board will respond to size press applications as measured by test like edgewick or internal bond. The internal bond or 'Z' direction strength is improved by size press starch only as far as penetration is achieved. To achieve edgewick or internal strength in the center of the board, internal wet-end applications must be used.

2. Normally cellulose fibers have many anionic sites exposed on their surface. Consequently, it is difficult to obtain a uniform distribution throughout the sheet when a highly cationic material is applied via a surface application. The highly cationic materials are retained near the surface by the anionic sites and do not penetrate as far as the liquid portion of the size press formulation.
3. The opacity of paper is largely due to the voids present in the sheet. Thus, any process which fills those voids will reduce the opacity. Consequently, to deposit substantial volumes of solids into a sheet via size press applications will lower the opacity of the sheet. This loss of opacity can be minimized by adding fillers to the formulation of reducing solids picked up. This effect is an advantage on those grades where reduced opacity is desired.
4. Some of the materials utilized in the size press formulations adversely affect the performance of other components in the sheet. For example, if the pH of the size press formulation is adjusted too high with caustic, the internal rosin size can be attacked and the water resistance substantially reduced. Size press defoamers are frequently such effective surfactants that they too will reduce the overall water resistance of the sheet, and if poorly dispersed, will even cause pinhole failures to penetrating liquids.
5. Size press applications contribute to high energy consumption since the water applied, which can be substantial, must be removed from a sheet that has already been dried once.

These few positives and negatives suffice to point out that the use of the size press is neither a utopia or a scourge, but is an alternative addition point that should be broadly considered in the design of each grade of paper or board. Size press utilization can be very simple, in fact, often simpler than attempting to obtain the same property via a wet-end application. When properly operated and the key variables controlled, it is a very efficient adjunct to papermaking.

References

1. R. W. Hoyland, B.Sc. Ph.D. 1971. *Basic Mechanism of Size Press Treatment of Paper*, University of Manchester, Institute of Science & Technology.
2. D. C. Kirk, Jr. Some Quantitative Studies in Liquid Pick-up at the Horizontal Size Press. *Southern Pulp & Paper Manufacturer*. August 10, 1967. 84-85.
3. J. A. Hansen and C. P. Klass. "High Speed Surface Sizing." 1983 Papermakers Conference Notes. 49-54. Atlanta: TAPPI PRESS, 1983.
4. D. R. Dill and K. A. Pollart. "What to Expect from Size Press Sizing." 1984 Papermakers Conference Notes. Atlanta: TAPPI PRESS, 1984.

5
Internal Sizing With Stearic Acid

T. Arnson, B. Crouse, W. Griggs

Introduction

The use of stearic acid as an internal sizing agent is limited to specialty papers with photographic papers being one of the principal applications. The method of size preparation and application is similar to rosin acids although its use is restricted due to its higher relative cost. Stearic acid provides excellent sizing resistance against aqueous solutions and, in contrast to rosin acids, will not oxidize thereby providing excellent brightness and whiteness stability.

Properties of commercial stearic acid

Saturated monocarboxylic fatty acids for the paper industry are available in chain lengths of C_8, caprylic acid, to C_{22}, behenic acid. These products have low iodine numbers and excellent color stability. The stearic acid grades are by far the most commonly used for paper sizing purposes. They are available in both flake and powder forms. As such, they appear as waxy, white, crystalline solids at room temperature. Stearic acid is a derivative of natural products, the source of which is found in the related word stearin, an ester of glycerol and stearic acid which is a common component in many animal and vegetable fats. The related French word "stearine" and Greek work "stear" both mean tallow or suet, a source material for stearic acid.

Commercial stearic acids are fractionated for specific end use applications. The actual fatty acid composition for a product sold as "stearic acid" may cover the range of values shown in Table 5.1. A typical papermaking grade of stearic acid would have the properties shown in Table 5.2.

Most of the stearic acids have limited solubilities in hot organic solvents. Those with 55% to 65% stearic acid, for example, may have 70% to 85% solubility at 50°C in such solvents as isopropyl alcohol, acetone, trichloroethylene and kerosene.

Table 5.1 Ranges of fatty acid composition in commercial "stearic acid."

	Fatty Acid	*% of Total*
$C_{14,}$	Myristic	0 to 5
$C_{16,}$	Palmitic	3 to 40
$C_{18,}$	Stearic	55 to 96
$C_{20,}$	Arachidic	0 to 3

Table 5.2 Typical properties of commercial stearic acid.

Iodine Number[1]	1
Saponification Number [2] (Acid No.)	201 to 207
Titer[3], °C	58 to 61
Molecular Weight	275

[1] Grams of iodine absorbed per 100 grams of fatty acid. A measure of unsaturation.
[2] Mg of KOH required to completely hydrolyze (saponify) 1 gram of material. Saponification number is inversely related to molecular weight.
[3] Solidification point.

Preparation and delivery of stearic acid for sizing

Higher order fatty acids, like stearic acid, are essentially water insoluble at room temperature and as such are unsuitable for addition to the papermaking system because of the poor mixing and distribution that would occur. However, at sufficiently high concentration, temperature, and pH, the stearic acid will exist as micelles and behave like a free-flowing solution.

Stearic acid is commonly made up at 2% to 3% total solids yielding an effective concentration far in excess of the critical micelle concentration of 1×10^{-3} M (1). The acid must be neutralized with an equivalent amount of sodium hydroxide although more sodium hydroxide can be added for the purpose of altering the pH of the stock at the point of addition. The solidification temperature of stearic acid is 60° to 65°C and, therefore, requires a higher make-up temperature, usually 80° to 85°C. The water and sodium hydroxide are mixed and brought up to temperature before adding the stearic acid. After 20 to 30 minutes of mixing, the hot sodium stearate is ready for delivery to the papermaking system.

The elevated solidification temperature also necessitates close monitoring of the delivery system temperatures in order to prevent sodium stearate aggregates. The temperature of the delivery solution, typically 70° to 75°C, must be maintained until the addition to the stock stream. Electrically traced or well insulated delivery lines are preferred as are smooth, uniform diameter pipes.

The best delivery results may be obtained by minimizing sharp angles in the delivery lines and restricting the number of blind side valves which can provide cool pockets in which size can accumulate and solidify. The elevated make-up and delivery temperature is a safety concern. Operators working with the make-up and delivery systems must wear the necessary protective clothing.

Development of sizing with stearic acid

The development of sizing with an aluminum stearate system generally follows the principles and concepts of alum-rosin sizing. The critical steps in the sizing mechanism follow from Guide's work on rosin sizing (2). These critical steps are: (1) formation of potentially low free surface energy precipitates, (2) deposition of the precipitates onto the pulp fibers, and (3) conversion of the wet size precipitate-fiber surface to a stable low free energy surface.

The addition of the sodium stearate into the stock system reduces the effective solution concentration of the stearate below the critical micelle concentrations (CMC), thereby forming the anionic stearate precipitate particle.

The particles range in size from 0.10 to 0.25 microns depending on the ionic environment. The sodium stearate should be added to the stock at a point of high turbulence in order to maximize the distribution of precipitate particles throughout the stock.

The addition of an aluminum salt converts the anionic sodium stearate to a cationic particle, aluminum stearate. The latter can readily absorb onto the anionic pulp fibers. The amount and form of aluminum will depend on the addition rate of stearate and on the types and amounts of other additives used in the wet-end system. In general, a minimum of 1:1 Al:stearate (molar basis) is needed to anchor the stearate to the fibers. The presence of additional anionic materials such as dry strength resins or retention aids may, however, require additional aluminum. Sizing can be obtained over a pH range of 4 to 5.5, and as pH controls the type of aluminum species formed, successful aluminum stearate sizing can occur when the aluminum exists as the simple trivalent cation, complex hydrolyzed species, or the colloidal precipitate. The differences in sizing response due to the aluminum chemistry and the overall chemical system is illustrated with data in Table 5.3. With the simple sizing System A, the aluminum precipitate (pH 5.2) has a better sizing response than the trivalent species (pH 4.2). The increased level of aluminum also improves the response in this system. On the other hand, when the system is further complicated with other wet- end addenda, the trivalent cation appears to be the preferred form of aluminum and the aluminum addition rate has only a minor effect. The increased sizing response in System B is due to an increase in fines retention.

Direct or reverse sizing can be used with the aluminum stearate sizing system. Direct sizing (stearate size followed by aluminum salt) generally yields a better sizing response and requires a stock pH of 7 to 8 before the addition of the sodium stearate. For mills with hard water, reverse sizing (aluminum salt followed by stearate size) is the system of choice in order to prevent calcium stearate deposits.

The final conversion of the aluminum-stearate particle to a stable low free energy surface requires heat and moisture in order to melt and distribute the stearate on the fiber surface. The effect of drying temperature and sheet moisture content on sizing response is shown in Figure 5.1. Below 40% to 45% sheet moisture, apparently there is insufficient water vapor to distribute the stearate properly (3). The 82°C to 88°C sheet temperature and 45% to 60% moisture content needed for maximum sizing response can normally be obtained with proper control of the initial drying zones of the paper machine.

Another aspect of the distribution of the stearic acid, which has yet to fully be quantified, is the vapor phase transport of the stearate during roll storage.

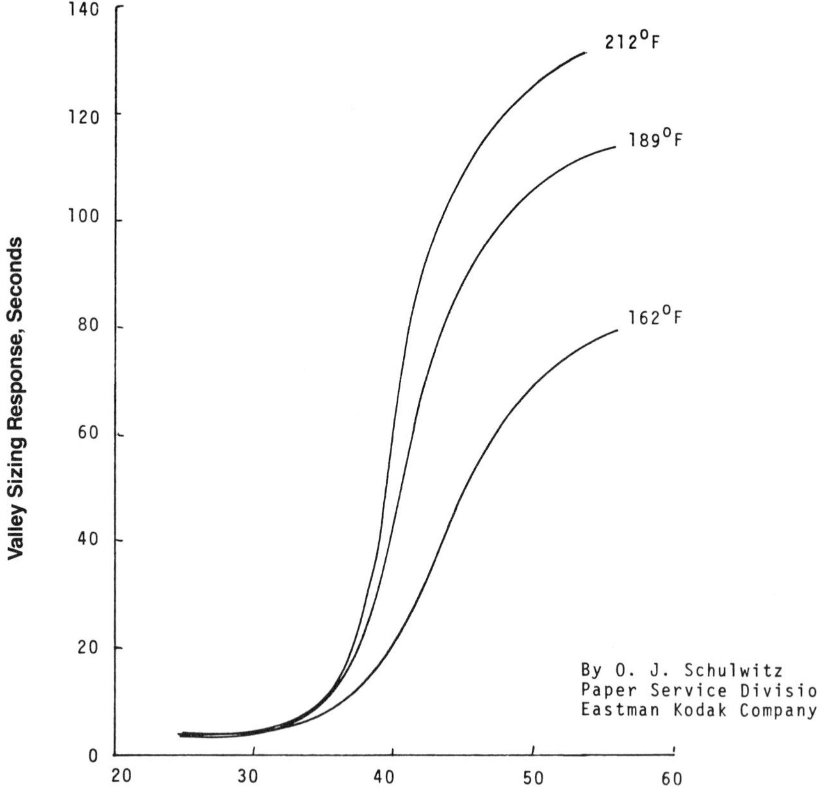

Figure 5.1 Effect of sheet moisture and temperature on stearic acid sizing.

The gradual loss of absorbency of absorbent products and the self-sizing of corrugating medium are known to be caused by the migration of residual resin and fatty acids (4). In addition, Swanson (5) has studied the mechanism by which cellulose fibers can be sized by adsorption of fatty acid from the vapor phase. However, the amount of stearate redistributed by vapor phase transport in a highly sized stearate paper may not be significant enough to change the apparent sizing response.

Table 5.3 Effect of aluminum chemistry on aluminum-stearate sizing.

pH	Aluminum Addition Rate, kg/metric ton	Valley Penetration, Sec	
		System A	*System B*
4.2	6.5	85	300
	13	126	290
5.2	6.5	145	250
	13	190	210

System A - Reverse sizing - Al, 9 kg/ton Stearic Acid
System B - Reverse sizing - Al, 9 kg/ton Stearic Acid, cationic wet strength resin, anionic dry strength resin

Sizing response with stearic acid size

The sizing resistance of stearic acid size is generally comparable to that obtained with rosin sizes, although it should be remembered that the particular penetrant and sizing test used also affect the measured sizing response (6). A comparison of a fortified rosin size and a commercial stearic acid is shown in Table 5.4. The measured response from the Cobb and Hercules Size Test indicate that both materials are similar in their sizing resistance, while the stearic acid demonstrates a superior resistance to acetic acid as measured by the Valley penetration test. The evaluation of stearic acid as an alternative sizing material should be conducted with the penetrant and size test that best correlates to the end use requirements of product.

The sizing resistance of fatty acid sizing materials is dependent on fatty acid chain length and generally increases as the chain length increases (7). Swanson (5) showed that behenic acid (C_{22}) was more efficient than stearate or isostearate. The data in Table 5.5 also show this trend although at a 50/50 mixture of palmitic and stearic, the influence of chain length is minimized. As the distribution of fatty acid varies for each grade of commercial stearic acid, the influence of chain length should be considered in the cost/performance of the stearate sizing material.

Table 5.4 Comparison of stearic and rosin acid sizes.

Size	Addition Rate, kg/metric ton	COBB gm/100 cm²	Valley² Penetration, Sec	Hercules Size³ Test, Sec
Stearic	6	.214	380	180
	12	.209	415	215
	20	.204	505	365
Rosin[1]	6	.222	8	220
	12	.210	10	510
	20	.203	12	915

[1]Pexol E400
[2]62.5% Acetic Acid
[3]40% Formic Acid

Table 5.5 Effect of fatty acid chain length on sizing response.

Fatty Acid	Valley Penetration, Sec[1]
Myristic (C_{14})	175
Palmitic (C_{16})	285
Palmitic/Stearic (50/50)	380
Stearic (C_{18})	385

[1]Penetration - 62.5% Acetic Acid

Mill problems with stearic sizes

Stearate spots

The formation of stearate spots in the finished product can occur due to both physical and chemical phenomenon. In either case, the stearate aggregates go through the wet-end system unretained by the cellulose and emerge in the white water or remain trapped in the sheet structure. Once calendered, the latter assumes the translucent appearance or other densified chemical deposits.

The elevated solidification temperature of sodium stearate lends itself to formation of aggregates if temperatures in the make-up and delivery system are not carefully monitored. Untraced or uninsulated delivery lines with valved recirculation or other side lines are particularly troublesome. Good housekeeping and proper maintenance can minimize the physical formation of stearate spots.

If an environment deficient in aluminum ion exists, the precipitation of aluminum stearate onto cellulose will be less than complete. Anionic polyacrylamide dry strength resins will compete with stearate for aluminum. Cationic additives, such as some of the neutral cure wet strength resins which are self-reactive with cellulose, may also set up a competition for available reactive sites on cellulose.

If this occurs in excess, precipitates of oppositely charged addenda and free stearate may appear in the system. These may then manifest themselves as objectionable spots in the product.

Wet press roll adhesion and stearate plating

Internal chemical systems that are deficient in aluminum ion may have the same side effects as a system with excess pitch, for example, press roll picking. The non-complexed free fatty acid may have a greater affinity for the press roll surface than it does for the fiber surface. If this occurs, the unbound stearate transfers to the roll surface, and redeposits on the paper surface when the press roll film becomes continuous and aggregates of free stearate begin to form. Efforts have been made to counter this by changing the press roll surface tension by substituting a press roll of different surface composition or by adding a surface active compound that will change the roll surface tension. The most direct remedy is to employ an adequate aluminum dose or reduce the stearate level to eliminate the excess free stearate.

Vapor phase migration of stearate

The vapor phase movement of stearate may be beneficial with respect to sizing response. This can be seen in some of the accompanying data that show the sizing response enhancement which is believed to be due to the more uniform distribution of the size after vapor phase movement. A liability of the vapor phase transfer of stearates may be found in the evolution of stearate from the paper in hooded dryer sections. In extreme situations, the evolved stearate may condense and solidify on the inner hood surfaces and then when the accumulation becomes great enough, flake off the deposited surface and settle back on the paper web to be compressed into the sheet at the machine calender. These dryer hood deposits may be regulated by careful attention to dryer temperature profiling and by providing adequate aluminum to bond the stearate and minimize the levels of free stearate.

Strength loss

In many respects, 10 or 20 kg per metric ton of fatty acid is similar to an equivalent amount of filler. Although it is bonded to the cellulose through the aluminum link, the bulky fatty acid does nothing to strengthen the paper structure. Controlled comparative studies have shown that stearate addition can weaken the paper strength. This is generally overcome by employing an appropriate dry strength additive and increasing the degree of fiber refinement.

Acid system

A paper sizing system which employs stearic acid incurs the same liabilities as any other sizing system which requires an acid salt for its successful development. Although stearic acid has optical permanence advantages over rosin sizing, the acid aging affects the physical properties as it does in any other acid paper. The long range effects of acidity or alkalinity on extrusion coated paper permanence are unknown.

Preferred uses of stearic acid size

The primary advantage of stearic acid over rosin type sizes is that the saturated fatty acid chain is not susceptible to the destructive yellowing of oxidation as with rosin acids. The optical properties of a stearate sized sheet thus have a greater permanence. This consideration, comparable sizing resistance, and higher cost have directed the use of stearic acid sizes toward hard-sized specialty paper grades which require good optical stability. Such grades include photographic paper support, drafting and tracing papers, and high rag content bonds.

Literature cited

1. Klevens, J. 1948. *Journal Phys. Chem.* 52, 130.
2. Guide, R. G. 1959. *Tappi.* 42, 9:740.
3. Glover, G. F. December 1957. *Brit Paper and Paper Makers Association, Proc Tech Sec.* 38, 3:515; discussion 528-33.
4. Swanson, J. W. and Cordingly, S. 1959. Tappi. 42, 10:812.
5. Swanson, R. 1978. *Tappi.* 61, 7:77.
6. Pickard, K. J. and Smith, A. Canadian Patent 1077946, "Treatment of Cellulosic Materials," (December 11, 1974).

6

Fluorochemical Sizing

R. M. Chad and C. A. Schwartz

Introduction

Fluorochemical is a term used to describe a unique class of organic compounds in which a large percentage of carbon bonded hydrogen atoms are replaced by fluorine atoms. The replacement of hydrogen by fluorine provides a low energy material that is immiscible in both aqueous and hydrocarbon systems. The general structure of fluorochemicals (hereinafter referred to as FC's) is shown below:

$$CF_3 \, (CF_2)_x \, R$$

The functional group R can be altered to provide a vast number of products with different properties. If R is a cluorine atom, the compound is inert, and is neither soluble nor dispersible in aqueous or organic solvents. Anionic, cationic, or nonionic functional groups can be used to make an FC surfactant. A polymerizable functional group produces a monomer which may be polymerized with other materials to provide products with a wide range of properties.

Zisman and his co-workers discovered that fluorocarbon surfaces, such as polytetrafluoroethylene, are of very low energy (1). These surfaces are not wet by aqueous or organic solvents. The application of FC's to paper allows the papermaker to obtain a low surface energy paper with properties much like the surfaces studied by Zisman. FC's enable the papermaker to manufacture paper with an invisible chemical barrier to oils and solvents without affecting the porosity of paper as with film forming chemicals or high degrees of refining.

Oil resistant papers

When a drop of oil is placed on a "greaseproof" paper or a polyolefin-coated paper, the oil will not penetrate because a physical barrier exists. However, the oil will spread over the surface of the paper. Oil placed on FC-treated paper does not spread; a discrete contact angle develops between liquid and solid, thus indicating that the force of cohesion (liquid- liquid interaction) is greater than the force of adhesion (liquid-solid interaction) (2). Figure 6.1 lists contact angles which develop when a drop of corn oil is placed on various paper surfaces after 10 seconds. Strong adhesional forces maximize the contact area between the oil and paper surface in all cases except the FC-treated paper. The FC treatment minimizes the oil-paper contact area.

Paper type	Contact angle with corn oil (in air)
Glassine	26°
Parchment	16
Polyvinyl alcohol coated	23
Hydrocarbon treated	0
Greaseproof	14
Polyethylene coated	14
Fluorochemically treated	129

Figure 6.1

The amount a liquid, B, will spread on a solid surface, A, can be defined by the spreading coefficient (S) as described by Cooper and Nutall(3):

$$S = \gamma_A - (\gamma_{AB} + \gamma_B)$$

where γ_A is the surface energy of the solid surface, γ_B is the surface tension of the liquid, and γ_{AB} is the interfacial tension. For spreading to occur, S must be greater than 0; for non-spreading less than or equal to 0. A reduction in γ_A, and a high γ_{AB} increases the likelihood of obtaining a value of $S \leq 0$, a non-spreading situation. A drop of oil on a polyethylene surface yields the following results:

$$S = 31 - (29 + 1) + 1$$

Because S is greater than zero one could predict the spreading that occurred with the polyethylene coated paper in Figure 6.1.

By applying the same calculation to an FC-treated paper:

$$S = 23 - (29 + 16) + -22$$

The non-spreading of corn oil on FC-treated paper, as shown by the high contact angle in Figure 6.1, is explained by a negative spreading coefficient.

The spreading coefficient only describes the thermodynamic relationship of wetting and spreading, it does not address the molecular structure or the surfaces of the materials. Work by Goebel assists in describing these aspects for FC treated systems (4).

Types of FC paper sizes

Three types of FC's are available to the paper industry: 1) FC chrome complex, 2) FC phosphates, and 3) FC polymers. The FC polymers can be further classified as one of two types: a) those polymerized with hydrophilic comonomers or b) those polymerized with hydrophobic comonomers. Each of these materials offers resistance to various types of low surface tension fluids.

Figure 6.2 shows the contact angle of four different liquids on paper treated with the different FC types. The data indicates that all of the FC's can make paper oil (Nujol) repellent, but if the papermaker requires holdout of other materials, such as water, alcohol, or hydrocarbon solvents, certain FC's can be more advantageous than others.

	FC-Chrome Complex	*FC-Phosphate*	*FC-Polymer: Hydrophilic Co-Monomer*	*FC-Polymer Hydrophobic Co-Monomer*
Water	134°	0°	136°	105°
Isopropanol	0	0	80	75
n-Octane	0	71	75	0
Nujol	119	127	126	116

Figure 6.2 Contact angles.

The FC chrome complex offers both oil and water repellency when applied to paper. It is not repellent, however, to low surface tension polar and non-polar solvents such as isopropanol or n-octane. The proposed structure for an FC-chrome complex is shown in Figure 6.3. Trebilcock has proposed a mechanism for the attachment of these types of compounds to cellulose fibers (5).

The FC chrome complex is no longer widely used on paper for several reasons: 1) The FC chrome complex is sold as a dark green liquid, so that even at low addition levels a slight green tint can be noticeable on bleached papers; 2) the FC chrome complex is unstable at a pH above 4; and 3) the advent of FC-polymers, which can provide similar repellency characteristics.

Two types of FC phosphates are currently available (Figure 6.4). The FC phosphates are widely used in the paper industry because they are cleared by the Food and Drug Administration for direct food contact packaging. These materials provide excellent holdout of oils and hydrocarbon solvents but, due to the phosphate group (a hydrophilic group necessary for aqueous dispersion) they are not repellent to water or other polar solvents. The FC phosphates can be co-applied with other materials to yield a paper which is resistant to both oils and water. This will be discussed in a later section.

As previously mentioned, the FC polymers can be separated into two types, those polymerized with hydrophobic comonomers, and those polymerized with

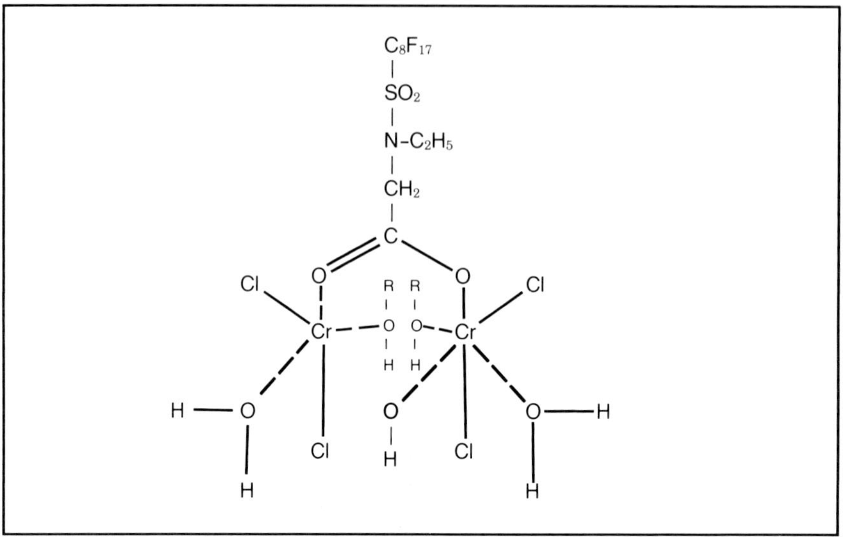

Figure 6.3 FC chrome complex structure.

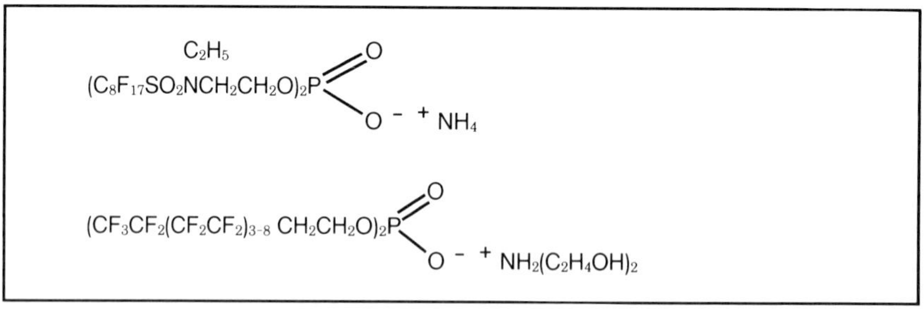

Figure 6.4 FC - phosphate structures.

hydrophilic comonomers. The general structures are shown in Figure 6.5. Under atmospheric conditions both types of FC-polymers provide resistance to water, polar solvents, and oils. So far, only FC- polymers with hydrophilic comonomers provide holdout of non-polar solvents.

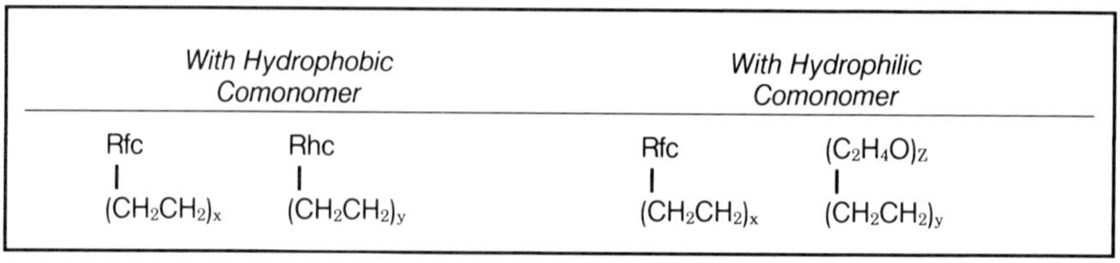

With Hydrophobic Comonomer		With Hydrophilic Comonomer	
Rfc	Rhc	Rfc	(C$_2$H$_4$O)$_z$
|	|	|	|
(CH$_2$CH$_2$)$_x$	(CH$_2$CH$_2$)$_y$	(CH$_2$CH$_2$)$_x$	(CH$_2$CH$_2$)$_y$

Figure 6.5 FC - polymer structures.

The FC polymers are used in many specialty paper applications; but because of their ability to hold out polar solvents such as alcohols, their primary impact has been made in the textile and medical nonwovens area (6).

Applications

One of the advantages of FC's is their application versatility. FC's can be applied by several techniques; internally, as surface additives at the size press or calendar stack; or in pigmented coatings.

Internal application (wet end)

FC's can be applied at the wet end of a paper machine. However, not all FC types are suitable for this method of application. Because of stability problems at a pH greater than 4.0, the FC chrome complex is not used in this type of application. The FC phosphates and FC polymers can be used internally, though the FC-polymers rarely are. Internal FC treatments can offer certain characteristics which would be difficult to obtain using conventional surface sizing techniques. By treating the individual fibers, the treatment is uniform throughout the sheet. This makes it less susceptible to physical abuse of the type encountered in calendering, supercalendering, or folding and creasing.

Since the FC phosphates, like cellulose, are anionic, a cationic retention aid is necessary. Usually, low molecular weight retention aids are suitable. The retention aid is usually added early in the stock preparation and is uniformly dispersed prior to the addition of the FC phosphate. The FC phosphates are sensitive to shear in internal systems, so they are generally added after screens and cleaners (7).

The phosphate group on the FC phosphates is unstable in the presence of multivalent cations, such as aluminum ions. Therefore, the FC phosphates are not used in acid sizing systems where the use of alum is popular. The combination of an FC phosphate and alkaline size, such as alkyl ketene dimer, can be used to achieve both oil and water resistance.

Both types of FC polymers can be added internally to a furnish to provide low levels of oil and water resistance. However, as Figures 6.6a and 6.6b indicate, these systems are not as efficient as a system utilizing the FC phosphate for oil resistance or the FC phosphate and alkaline size for oil and water resistance.

The FC treatment levels in Figures 6.6a and 6.6b range from 0.0% to 0.5% FC solids by weight of fiber solids (S/F). When the alkaline size (8) was co-applied with the FC phosphate a ratio of 5:1 (alkaline size: FC phosphate) was maintained based on S/F.

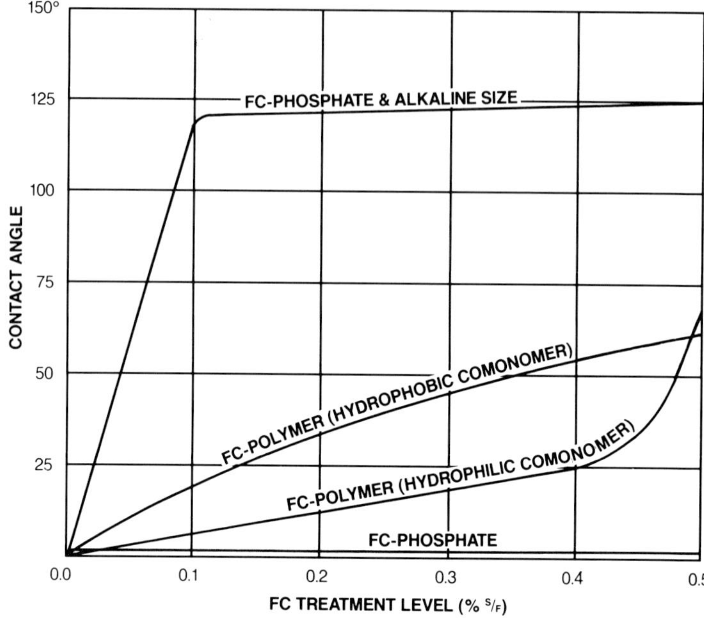

Figure 6.6A Oil resistance of internal FC treatments (contact angles with Nujol)

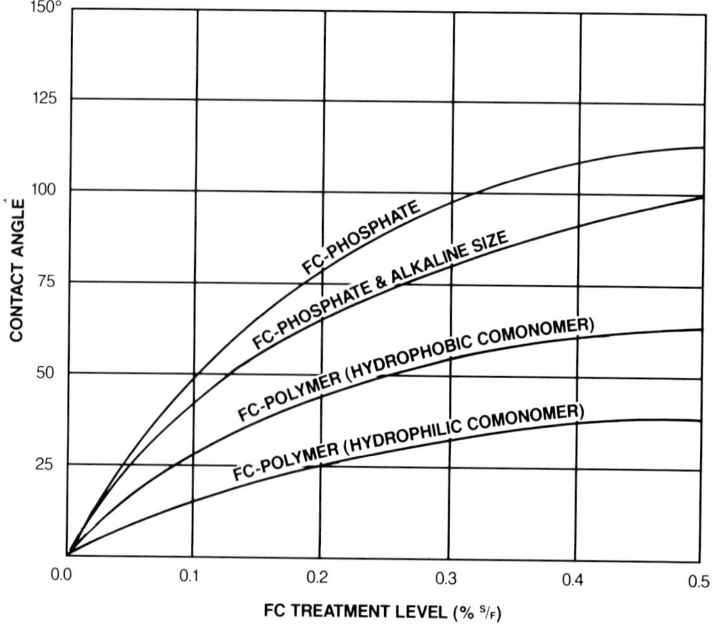

Figure 6.6B Water resistance of internal FC treatments (contact angles with water)

Size press application

The size press is the most common method of applying fluorochemicals. All of the different types of FC's can be applied by this technique. Size press treatments are usually more efficient than internal treatments from an economic standpoint.

The two types of size press applications offer either (1) maximum creased oil resistance or (2) flat oil resistance. Work by D. R. Dill has shown how to maximize size press pick-up and penetration (9). These principles can be used in applying FC's to obtain maximum creased oil resistance. The key factor in obtaining good depth of treatment at the size press is the elimination or near elimination of internal size. This, along with the removal of film forming material from the size press solution, such as starch or polyvinyl alcohol, can allow the FC to penetrate the web and provide an "internal-like" treatment.

The second type of size press treatment is that which yields flat or surface resistance to low surface tension fluids. This is achieved by co-applying the FC with some type of film former, usually starch or polyvinyl alcohol. As the level of film former increases at a constant FC treatment level, the degree of low surface tension fluid holdout also increases (10). This type of treatment is very susceptible to physical abuse. If the finished sheet is subjected to folds or creases, surface holdout may be destroyed.

Depending on the type of FC being applied at the size press, certain guidelines must be followed in solution preparation. As previously mentioned, solutions of the FC chrome complex are unstable at a pH greater than 4.0, therefore, pH control is essential. The FC chrome complex is unstable in the presence of multivalent anions. Generally, a carbonate hardness less than 200 parts per million can be tolerated without a loss in solution stability. During the FC chrome complex "curing" process, hydrochloric acid evolves. Urea is commonly added as an acid acceptor in the treating solution (11).

The FC phosphates are unstable in the presence of multivalent cations such as CA^{2+} and Al^{3+}. The addition of a suitable chelating agent prior to the addition of the FC phosphate alleviates potential problems by ensuring solution stability. The FC phosphates are stable over a fairly broad pH range (5 to 10), so pH control is usually not necessary (12).

The FC polymers can act quite differently in size press treating solutions. The FC polymers with hydrophilic comonomers are stable in solution over a broad pH range, usually 3 to 10. However, those with hydrophobic comonomers may be stable only at a narrow pH range (4.5 to 5.5). When an FC polymer is co-applied with starch, those with hydrophilic comonomers are more stable than those with hydrophobic comonomers (13).

Figures 6.7a and 6.7b show the oil and water repellency obtainable with different percent S/F levels of the various FC types. These treatments would be of the type to obtain maximum creased oil resistance. It should be noted that the FC phosphate offers no resistance to water. This is due to the very hydrophilic phosphate group. However, if an FC phosphate is co-applied with styrene-maleic anhydride resin (14) or a polyurethane size (15) a combination of oil and water repellency can result.

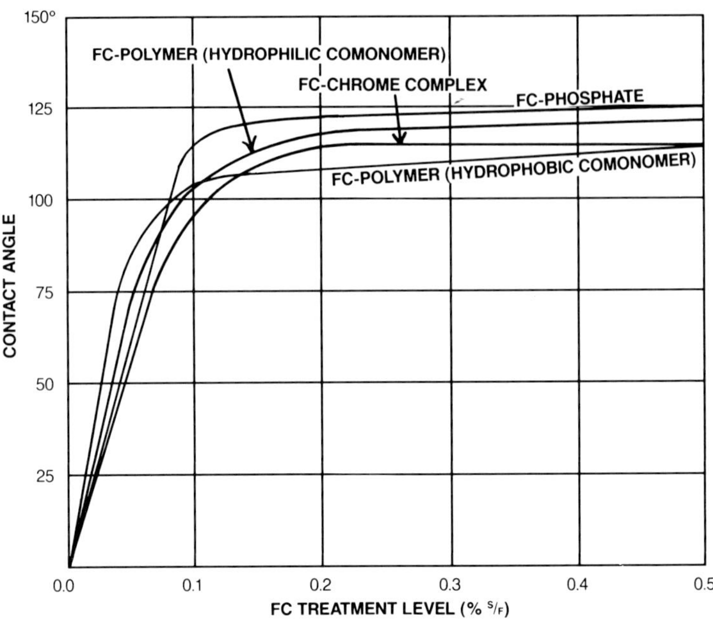

Figure 6.7a Oil resistance of FC size press treatments (contact angle with Nujol).

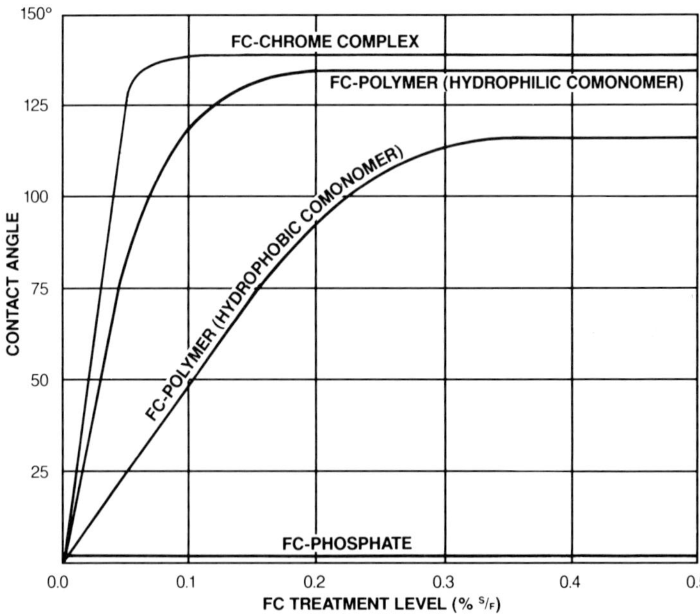

Figure 6.7b. Water resistance of FC size press treatments (contact angle with water).

Pigmented coating applications

Pigmented coatings have been treated with FC's to render them resistant to various low surface tension fluids. Here the FC phosphates are used almost exclusively. The FC chrome complex is not used because of stability problems due to pH. The FC polymers can be used to provide degrees of oil and water repellency, but their poor efficiency usually precludes their use.

The large effective surface area associated with pigmented coatings accounts for the increased level of FC necessary to provide practical repellency. Figures 6.8a shows that FC phosphates provide good oil repellency at a threshhold level of 1.2% to 1.5% FC solids by weight of total coating solids; the FC polymer with a hydrophilic comonomer never achieves the same level of oil repellency even at 2.0% solids on coating solids. The FC polymer can offer some water resistance beyond that which is inherent to the coating, while the FC phosphate offers no additional water holdout as seen in Figure 6.8b.

The coating used in Figures 8a and b has the following formula: 100 parts pigment, 10 parts protein, and 10 parts polyvinyl acetate. The oil resistance achieved can vary depending on the coating formula and the types of materials used. For instance, styrene butadiene latex binders are more oleophilic than polyvinyl acetate binders; therefore a higher level of FC phosphate may be necessary to obtain the same level of oil resistance.

Calender stack applications

FC phosphate calender stack treatments are common on paperboard. The solution formulation is very similar to size press applications for flat oil resistance. The FC phosphate is usually co-applied with a film former and applied out of two water boxes to ensure good uniformity.

Effect of cure temperature

The degree of oil repellency obtained with an FC treatment is related to how "closely packed" the perfluoroalkyl groups are at the substrates surface. This depends on both the density of the perfluoroalkyl groups, i.e. treatment level, and on their alignment. Elevated temperature curing is often required to provide sufficient mobility to allow the FC to move to its thermodynamically stable, low energy state (17).

The final orientation or alignment of the FC will depend on the substrate, the FC structure, and the application/drying environment. Most FC paper treatments are designed so that the perfluoroalkyl groups orient into the air interface and the functional groups or the hydrophilic polymer segments are associated with the cellulose (usually through hydrogen bonding).

A comparison of cure temperature versus kit rating (16) for two different types of FC polymers is presented in Figure 6.9. For this study "waterleaf"

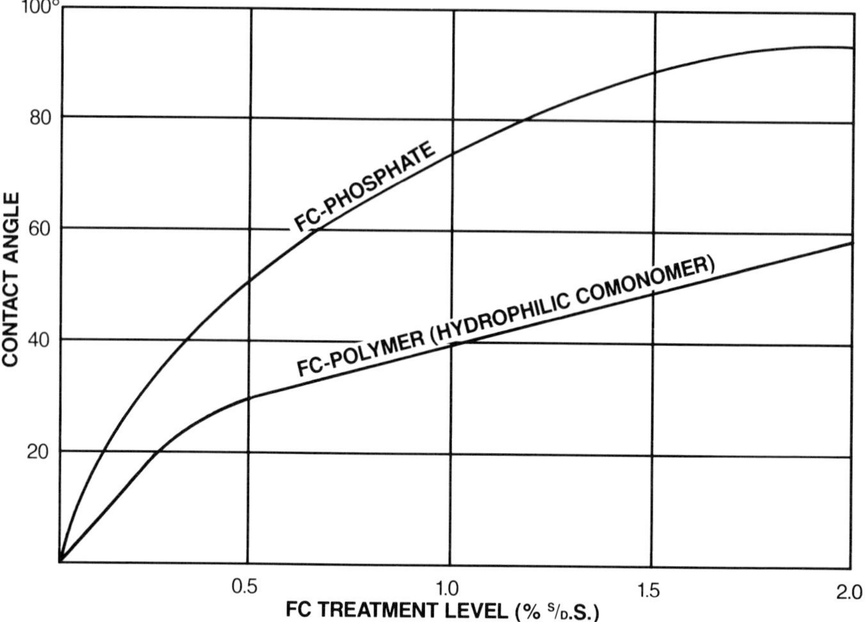

Figure 6.8a Oil resistance of an FC treated pigmented coating (contact angle with Nujol).

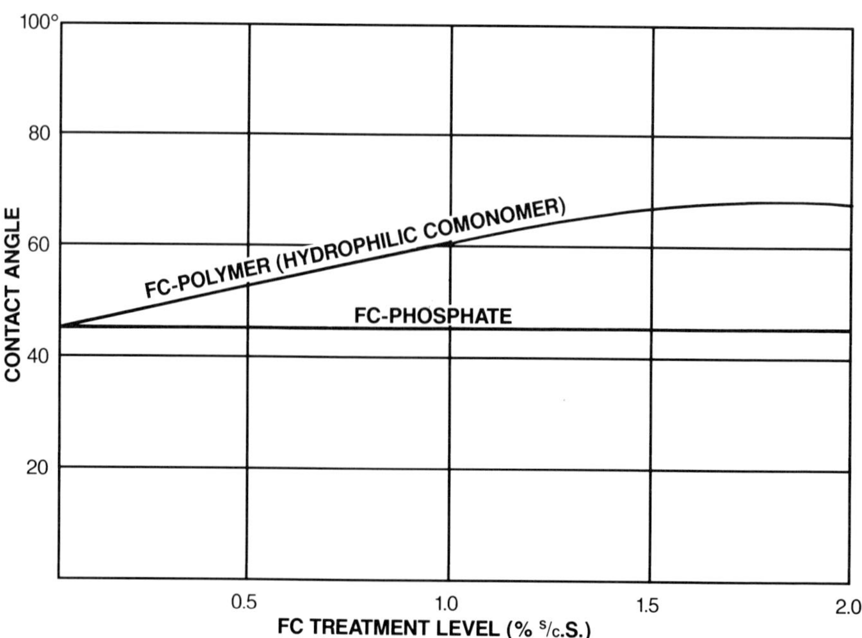

Figure 6.8b Water resistance of an FC treated pigmented coating (contact angle with water).

paper was treated at 0.1%, 0.2%, and 0.3% FC S/F. Each sheet was dried on a series of aluminum blocks which ranged in temperature from 65°C to 121°C. The test sheet was covered with a papermaking felt and dried for 30, 60, or 90 seconds. Block surface temperature, paper/felt interface temperatures, dwell times, treatment levels, and kit ratings for each block were recorded and compared. The results showed that the degree of oil resistance at each treatment level was dependent on the paper/felt interface temperature. The particular block surface tempera-tures and dwell times (over the range studied) required to reach the paper/felt interface temperature were not important.

Figure 6.9 shows that the FC polymer with the hydrophobic comonomer requires much higher curing conditions than the FC polymer with the hydrophilic comonomer. The FC polymer with the hydrophilic comonomer is usually preferred for most paper applications because of its lower cure requirements.

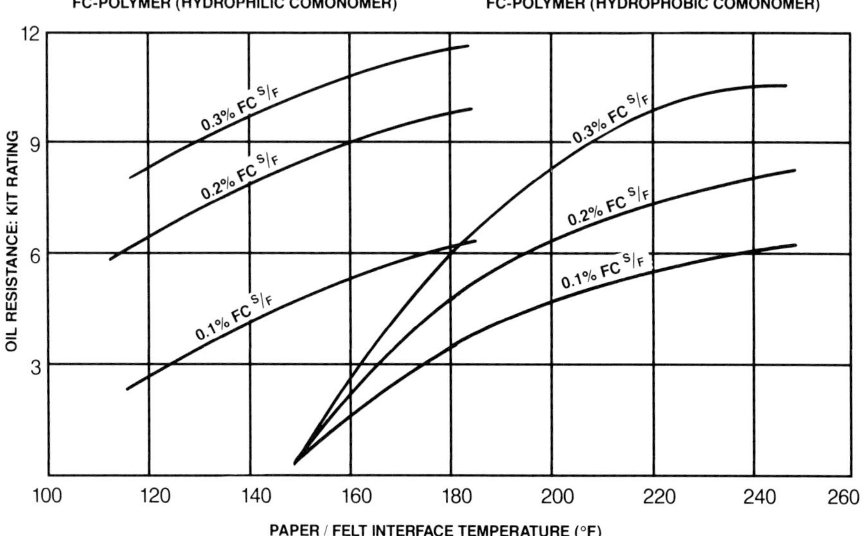

Figure 6.9 Oil resistance vs. cure temperature.

The other two types of FC's orient under very mild cure conditions. The FC chrome complex requires only slightly elevated cure conditions, with sheet temperatures in the range of 100° to 130°F. The FC phosphates do not require elevated cure temperatures. A sheet treated with an FC phosphate will yield equivalent performance whether it is dried at room temperature or under paper machine conditions.

Effect on physical properties

FC treatments generally have no effect on the odor, the appearance, or the "feel" of paper. Since they are chemical treatments and not film formers, they have no effect on porosity, moisture vapor transmission rate or the flexibility of paper. FC treatments have little, if any, effect on repulpability.

As mentioned earlier, however, the FC chrome complex may leave a green tint on bleached grades at high treatment levels. The FC chrome complex, if not properly applied, can also cause a reduction in strength due to the liberation of hydrochloric acid during curing.

The FC phosphates can cause a reduction in strength when they are used in size press treatments, particularly when no internal sizing is used. This effect is believed to be related to the low surface tension of the treating solution. When debonding occurs at the size press, the surface tension forces which draw the fibers together during drying are less with the FC phosphate treating solution (18). The loss of strength can be minimized by the addition of low levels (2%) of either starch or polyvinyl alcohol. No strength losses have been observed with the FC polymers.

Effects on machine operations

The fiber, fillers, and chemical additives all can affect FC treatments through physical or chemical factors. Physical factors include the nature of the surface area to be treated as determined by fiber type, refining, and the levels and types of fillers. Chemical factors include oleophilic materials such as pitch, lignin, rosinates, and hydrocarbon based additives, all of which decrease oil resistance.

For FC phosphate internal treatments, bleached pulps are easier to treat than unbleached pulps, sulfates easier than sulfites, softwood easier than hardwood, and chemical pulps easier than groundwood (19).

Fluorochemical treating solutions generally have a tendency to foam and stabilize foam. It is important to eliminate any foam generating sources such as free fall in return lines. Several effective defoamers have been identified which are added to the treating solution at 100 to 1000 parts per million prior to the addition of the FC.

Ordinary machine calendering has little or no effect on the repellent properties of FC treatments. However, supercalendering does decrease the surface performance of FC treatments. This effect is probably due to the exposure of untreated areas and the disruption of the FC alignment. Increasing smoothness also affects the contact angle at the solid/liquid interface (20). Supercalendering generally has no effect on the edge wick resistance of FC treated papers.

FC treatments can decrease the adhesion of extruded films. The FC phosphates and FC polymers with hydrophilic comonomers seem to have the greatest reduction in film adhesion. Corona discharge has been used to improve film adhesion of FC-treated substrates.

Due to the low surface energies of FC-treated papers, printability can be a problem, especially in offset printing where oil-based inks are used. Inks have been designed with lower surface tensions to provide better wettability which alleviates the problem. FC treatments can also interfere with drying rates of offset inks where ink absorption is important. Oftentimes, the level of driers in offset inks is increased to speed drying. Virtually no problems have occurred with flexographic or rotogravure printing.

Adhesives (especially hot melts) have been modified to provide better wettability and improved adhesion to the low surface energy FC-treated papers.

Testing FC-treated papers

Because of the wide variety of materials held out by FC paper treatments, specific end use testing is imperative. End use testing is used to determine the most suitable FC treatment level, treatment depth, and application method. This information is then correlated to simpler test methods to be used for quality control.

Useful FC treatment tests include those listed in Figure 6.10. Not all of these are standard tests, as modifications in testing are often necessary to fit a specific end use. There are three tests, however, that have gained fairly wide acceptance for FC-treated papers:

(1) TAPPI-UM 557 - "Repellency of Paper and Board to Grease, Oil, and Waxes (Kit Test)." This test involves oils numbered 1-12 which contain specific ratios of three reagents; castor oil, toluene, and heptane. Oil #1 is the least aggressive oil or the one with the highest surface energy, and oil #12 is the most aggressive oil or the one with the lowest surface energy. The various oils are dropped onto the substrate from a height of one inch. After 15 seconds the oils are removed with tissue, and the "Kit Rating" is defined as the highest numbered oil which does not stain the substrate. Therefore, the greater the "Kit Rating," the greater the degree of oil resistance.

Test	Measurable quantity	Reference
Product storage	Time to fail, % stain, % wt. gain, etc.	. . .
Oil wicking	Ht. of wick, brightness loss	. . .
Oil or wax dip	Brightness loss, % wt. gain	Tappi UM 478
Oil wipe	Pinhole frequency	. . .
Solvent holdout	% penetration	Tappi UM 528
Turpentine	Time to fail	Tappi UM 416
Alcohol Holdout	Std. rating scale	Inda 80.9
Cobb	Wt. gain	Tappi 441
Liquid penetration	Time to fail	HST*
*Hercules size test.		

Figure 6.10

(2) TAPPI-T458 - "Surface Wettability of Paper (Angle of Contact Method)." A drop of the liquid to be tested is placed on the paper surface. The contact angle is then measured through the liquid to a line drawn tangent at the point of contact of liquid, solid, and vapor (air) interface (21). This measure of wettability can be used to correlate degrees of repellency. An advantage of this test is that the liquid can be varied to closely simulate end use.

(3) Fluorine Analysis - This test method can provide an accurate determination of the amount of fluorine in the paper. With this information and a suitable conversion factor, the amount of the FC treatment on the paper is calculable. A sample of the paper is decomposed in an oxygen flask to convert fluorine to fluoride. The fluoride concentration can then be determined by conventional techniques, including titration, fluoride electrode, or a colorimetric method (22).

References

1. Zisman, W. A. 1964. *Adv. in Chem.* 43,1.
2. Stryker, L. J. 1983. *TAPPI 1983 Sizing Seminar Notes.* Atlanta: TAPPI PRESS.
3. Cooper, W. A. and Nuttall, W.A. 1915. *J. Agr. Sci.* 7,219.
4. Goebel, K. D. 56th Colloid and Surface Science Symposium, Tech. Presentation, 1982.
5. Trebilcock, J. W. 1981. *TAPPI Sizing Short Course Notes.* Atlanta: TAPPI PRESS.
6. Colbert, J. F. and May, M. A. 1981. *Index 81-Congress Papers.*
7. Scotchban® Paper Protector FC-807, 3M, Product Information.
8. Hercon® 40, Hercules Inc., Wilmington, DE.
9. Dill, D. R. 1974. *TAPPI.* 51 (1), 97.
10. Schwartz, C. A. 1981. *TAPPI Sizing Short Course Notes.* Atlanta: TAPPI PRESS.
11. Scotchban® Paper Protector FC-805, 3M, Product Information.
12. Scotchban® Paper Protector FC-807, 3M, Product Information, and Zonyl® RP, E. I. duPont de Nemours and Co., Inc., Product Information.
13. Scotchban® Paper Protectors FC-808, FC-829, FC-824, 3M, Product Information.
14. Scripset® 720, Monsanto Plastics and Resins Co., St. Louis, MO.
15. GrapHsize® A, Armak Co., Maple Shade, NJ.
16. TAPPI UM 557
17. Sherman, P. O., Smith, S. and Johanessen, B. 1969. *Textile Res. J.* 39, 441.
18. Campbell, W. B. 1933. Can. Dept. Int., Forest Service Bull. No. 84.
19. Rengel, G. L. and Young, R. C. 1971. *TAPPI Monograph Series No. 33.* Atlanta: TAPPI PRESS.

20. Dettre, R. H. and Johnson, R. E., Jr. 1964. Contact Angle, Wettability, and Adhesion. *Adv. in Chem. Series* 43.
21. Stryker, L. J. 1983. *TAPPI Sizing Seminar Notes.* Atlanta: TAPPI PRESS.
22. Schwartz, C. A. 1981. *TAPPI Sizing Short Course Notes.* Atlanta: TAPPI PRESS.

7

Testing Paper and Board for Sizing

R. W. Kumler
J. M. Gess

Introduction

All papers and boards will sorb liquids. This chapter is concerned with the sorption of aqueous liquids by paper and board.

At this time, there is no one universal method for measuring the sorption of water by paper and board. As a result, an understanding of the types and mechanisms of water sorption by paper and board and a knowledge of what parameters the tests available for this measurement cover and what the paper or board is to be used for is a critical guide in selecting a given test.

This chapter will concern itself with what is known of the measurement of the sorption of water by paper and board. The choice of what test to use, which will be shown, is a function of many factors. Thus, this chapter cannot make a specific recommendation as to what test to use, nor can one test be recommended as "the" test for sizing in paper.

Mechanism of the sorption of water by paper and board

The capacity of paper to retard penetration of liquids is known as "degree of sizing" or "size resistance." Carson (1) points out eight ways that water, contacting one side of the sheet only, may penetrate paper:

Entering Phase	Migrating Phase	Emerging Phase
1. Liquid	Liquid	Liquid
2. Liquid	Liquid	Vapor
3. Liquid	Vapor	Liquid
4. Liquid	Vapor	Vapor
5. Vapor	Liquid	Liquid
6. Vapor	Liquid	Vapor
7. Vapor	Vapor	Liquid
8. Vapor	Vapor	Vapor

We are here concerned primarily with the first case, but any of the first four may describe thesituation at some stage of the transudation of the liquid. The entering phase is subject tocontrol, and distinction can be made between liquid and vapor in the emerging phase. If theentering phase is liquid and emergence of liquid only is noted, case 1 is probably what is beingmeasured. However, if there is considerable temperature difference between the two sides of the paper or board being tested, case 3 could be the method of transudation. When the liquid is held at room temperature or below, cases would exist in the following order: case 4, case 2, case 1. Under conditions of rapid evaporation from the free surface of the sample, case 1 may not appear at all, at least within any reasonable time limit.

Paper, by nature, is porous. The solid fraction of paper may occur anywhere between about 15% and 85% of the total volume, depending on the grade (2,3). The air fraction of paper is classified as recesses, pores, and voids. Recesses are surface indentations that do not extend through the sheet. Pores are channels or openings that extend from one surface of the sheet to the other. Voids are spaces not connected to either surface. The effect of voids on sizing measurement has not really been thoroughly investigated. Pores are irregular and multibranched, many running roughly parallel to the plane of the paper. The internal sizing of paper does not appreciably fill the interstices between the fibers nor does it completely cover the fibers. In a well-sized sheet the particles are sufficiently close together to raise the overall contact angle and thereby retard the penetration of aqueous liquids without effectively impeding the passage of moisture vapor.

Testing paper for sizing

Wilson (4) suggests that there are two paths by which aqueous liquids can pass through paper, namely, via the capillaries or interstices between fibers (intrafibrillar penetration), and via the cellulose structure (intrafibrillar penetration) itself. Since sizing is primarily a treatment of the surfaces of the fibers, intrafiber penetration should not be affected by it. Verhoeff and coworkers (5) concluded that water penetrates paper by the two methods suggested above and that the rates of penetration by the two methods are controlled by different factors. For instance, the rate of flow through interfiber pores varies inversely with the viscosity and/or surface tension of the test liquid, whereas intrafiber penetration is independent of viscosity when a solution of carboxymethyl cellulose is used as the test liquid. This is explained by the premise that the carboxymethyl cellulose molecules are too large to enter the cellulose structure and only the water goes through. Sizing degree was determined by the Valley Test Method. Verhoeff and coworkers (5) present evidence that interfiber bonds form pathways for the passage of water. It is therefore not necessary that individual fibers extend from one surface of the sheet to the other in order that this type of penetration can occur.

Reaville and Hine (47) have examined the mechanism of water penetration into papers sized with a variety of sizing agents. Their data support the conclusion that the primary method of penetration is by surface diffusion through the pores of the paper followed later by rapid penetration into the fiber. However, this would depend on such factors as the average diameters of the pores in the sheet being tested; the surface tension and viscosity of the test fluid being used (and the hydrostatic pressure; i.e., test head, on the system) and the degree of sizing of the fibers.

Further, Reaville and Hine's conclusions do not hold for papers and boards that are surface treated. As shown by the work of Gess (49) in Figure 7.1, surface sizing can have a marked effect on the initial rate of penetration of water into paper. One group of tests - the Wet Break, Penescope, and ink Flotation Tests - appears to measure properties related to penetration of water into the fibers. A second group of tests - the Valley, KBB, and Fluorescence Size Tests - appears to measure some function of the transudation rate of water through the pores. A third group of tests - the Cobb, Curl, and Water Pickup Tests - measures some function of both mechanisms of water penetration.

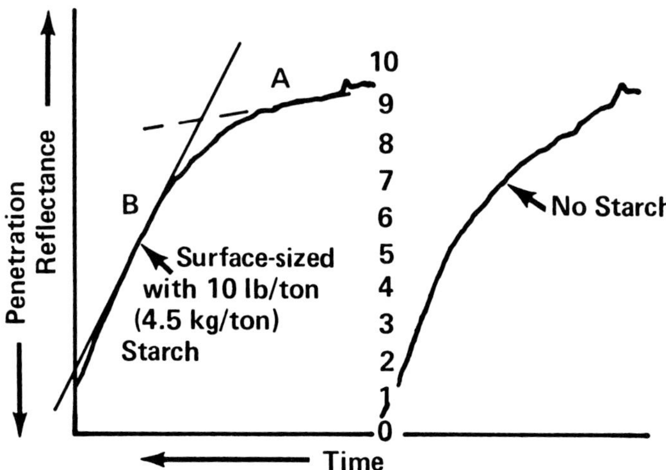

Figure 7.1 Effect of surface sizing on the rate of penetration of fluids into paper and board

The Washburn equation for the flow of liquid through a capillary is as follows (6):

$$\int^2 = \frac{r\gamma \cos\theta\, t}{2\eta} \tag{1}$$

where

$\int$ = depth of liquid penetration, cm
r   = pore radiation

γ = surface tension of the liquid, dynes/cm
θ = angle made by contact of liquid with solid
t   = time of penetration, sec
η = coefficient of viscosity of the liquid, poises

Cobb (6), who introduced Eq. 1 as applied to sized paper, pointed out that its chief value lies in clearly indicating the factors that determine the rate of penetration into an empty paper capillary. These factors may be summed up as follows.

At constant temperature the depth of penetration of an aqueous liquid into paper varies with the following factors:

1. Directly as the square root of time t
2. Directly as the square root of pore radius r
3. Directly as the square root of surface tension γ
4. Directly as the square root of the cosine contact angle θ
5. Inversely as the square root of the viscosity of the test liquid η

Several facts must be kept in mind when interpreting and applying the Washburn equation.

1. The equation is applicable only to employ capillaries and is not valid for the rate of passage of water through paper after the capillaries are filled.
2. The equation is applicable only when the contact angle with the test liquid is acute. When the contact angle exceeds $90°$, cosine θ becomes negative.
3. The effect of the surface tension is understandable only when it is borne in mind that surface tension influences the contact angle. For unsized papers where the contact angle is zero or nearly so, the depth of penetration varies directly as the square root of the surface tension. However, for sized papers a reduction of the surface tension of the test liquid may so affect the contact angle that depth or rate of penetration is increased.
4. As applied to testing paper for water resistance, this equation really has no constants. As paper is exposed to aqueous liquids, swelling takes place, which changes the pore radius and the contact angle. In most cases the test liquid becomes slightly contaminated by substances leached from the sheet and its characteristics are altered. Nevertheless, the equation is useful as a means of expressing the influences of various factors in making size tests.

Van den Akker and Wink (48) more recently compared the permeation of papers as measured by the Fluorescence Size Tester with that measured by means of an early rate method used for measuring surface receptivity of papers. Their considerations led to the conclusion that permeating speed is nearly constant and that size time is nearly independent of the viscosity of the permeating solution.

It was shown (49) that in agreement with the work of Van den Akker and Wink (48), the rate of penetration of an aqueous test fluid into a handsheet is linear. The slope of the rate of penetration vs. time (as measured using the list) depends on the degree of sizing within the sheet being tested. Further, the work (49) has shown that the surface treatment of the web (i.e., by a size press) gives a rate of penetration vs. time plot such as that shown in Figure 5.1c. Part A of the plot is the rate of penetration due to the surface treatment. Part B of the plot is due to the internal structure of the web and the hydrophobicity of the fibers.

The so-called "gate" effect should also be noted at this point. When the test fluid cannot dissolve the surface treatment (i.e., where the surface treatment is a polymer), the rate of penetration of the test fluid (Part B) does not have to be a function of the internal structure of the web and the hydrophobicity. Rather, the rate of penetration is a function of the rate at which the test fluid is being allowed to enter the web (i.e., the surface treatment is acting as a "gate" and controls the rate of penetration of the test fluid into the sheet). This is shown in Figure 7.2.

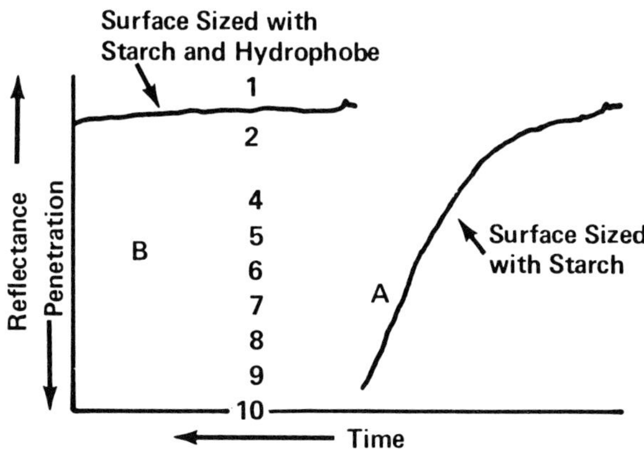

Figure 7.2 The "gate" effect. Control of penetration via the pores of a nonsoluble polymer on the surface of the paper.

Factors affecting testing

A number of factors affect the rate of penetration of aqueous liquids into paper or paperboard independent of the degree of the sizing of the paper. These factors assume variable importance according to the type of test being conducted, but several can be listed as being influential in a majority of tests. The most important are as follows:

1. Temperature of sample, testing liquid, and atmosphere in which test is conducted

2. Moisture content of sample
3. Pressure of liquid against the sample
4. Composition of testing liquid
5. Preparation of the sample, including wrinkling, creasing, soiling, etc.

Other factors may appear in connection with specific test procedures.

Effect of temperature

The higher the temperature of the test liquid, the more rapidly it penetrates paper or board. Therefore it is very important to control temperature. Libby and Casciani (7) studied the effect of temperature in connection with the testing of paperboard with lactic acid in the Penescope. Varying the temperature of the lactic acid from approximately 7°C to 37°C caused changes in penetration times from 176 to 9 min. in one case, and from 140 to 11 min. in another. Significantly, the effect of a 0.5°C change was greater in penetration time in the cooler range than in the hotter range. Based on one typical temperature-penetration curve the following data are noteworthy.

Table 7.1 Variation of penetration of lactic acid with temperature.

Temperature Range, °C	Penetration Time Range, min.	Change in Penetration per 0.5 °C
25 to 37	24 to 9	0.7
17 to 25	56 to 24	2.3
6 to 17	176 to 56	6.3

These figures may not be precisely applicable to other samples tested by the same method or to other test methods, but it is safe to assume that temperature would be as important in any procedure when time of penetration of aqueous liquid into or through a sheet is being measured. Van den Akker, Nolan, Dreshfield, and Heller (9), reporting on studies of the Fluorescence Size Tester, found that penetration times varied about 5% per °C change of the testing liquid near room temperature.

Codwise (10) has studied testing for water transudation time by floating the sample on water at temperatures up to the boiling point. At water temperatures down to 22°C he confirmed thegeneral findings of Libby and Casciani (7). He further suggested that for a given set of sizing conditions it is possible to establish temperature-penetration-time curves by which hard–sized paper or board may be tested relatively quickly at temperatures near the boiling point, and the data translated into penetration times at lower temperatures.

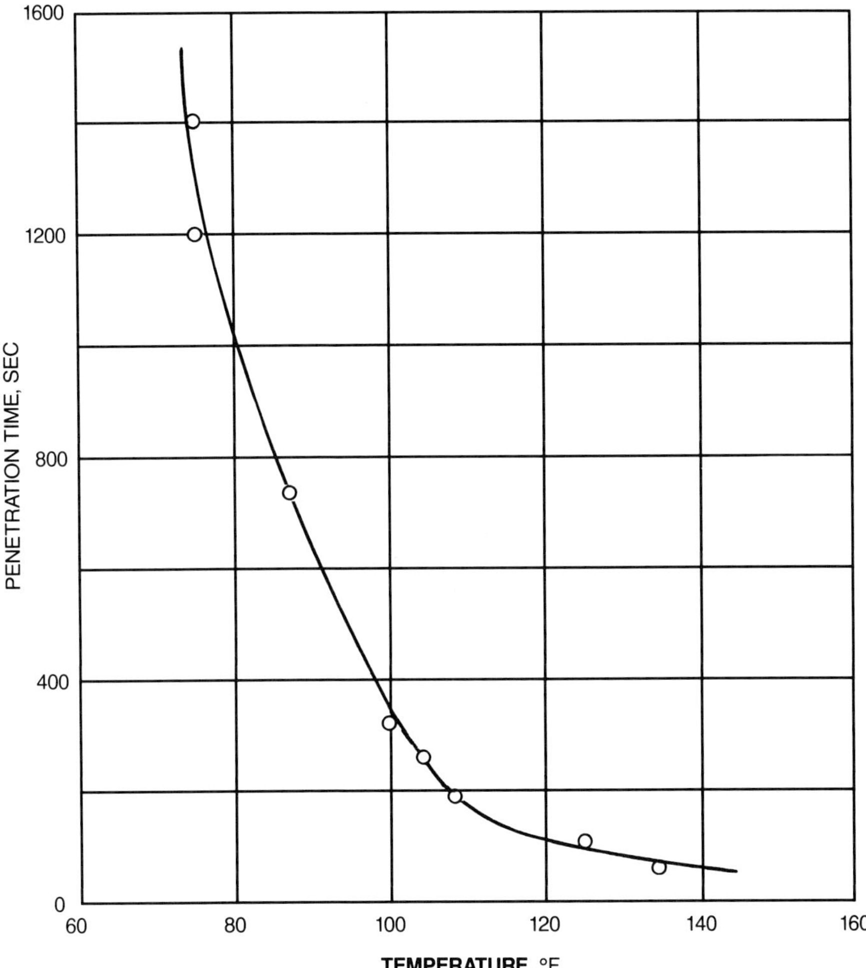

Figure 7.3 Effect of temperature on lactic acid Penescope Test (17 point milk carton stock).

In attempting to translate test results obtained at high temperatures into test values at room temperature, care must be taken to establish the relationship by experiment. It should be recalled that wax-containing sizes impart excellent resistance to water at 70°F but are ineffective against very hot water. It is possible that a study of the subject might reveal other factors that affect the relationship between high- and low-temperature tests. Of course, where the product being tested is designed for use with hot liquids, such as coffee, high–temperature tests are required.

Effect of moisture content of paper

The moisture content of the sample has an effect on the rate of penetration of water. Codwise (11) noted that by the dry indicator test transudation time was 22.7% higher when the sample was desiccated over calcium chloride than when conditioned at 65% relative humidity. Van den Akker and coworkers (9) showed that an average increase of 1% relative humidity in which the sample was conditioned reduced the size time 0.39 sec. This points up the importance of conditioning samples carefully prior to testing. The TAPPI procedure for conditioning samples is given in T 402.

Effect of hydraulic head

The effect of pressure on the rate of penetration of an aqueous liquid into or through a sheet of paper on the Penescope was studied by Libby and Casciani (7). When a 0.020-in. bleached lined manila board was tested with lactic acid, penetration was instantaneous with a 23-in. head but took 32 min. with zero head. Instantaneous penetration was noted for 0.040 bottle-cap board at 28 in. of head, and 45 min. was required at zero head.

Table 7.2 (8) shows the effect of hydraulic head on lactic acid penetration times in the Penescope.

Table 7.2 The effect of hydraulic head on lactic acid penetration time in the penescope.

Hydrostatic head, in.	0	6	12	22
Penetration time, sec	3250	2510	2050	1200

Composition of the test liquid

It was noted above that the rate of penetration of a liquid into paper depends in part on the surface tension and viscosity of the liquid. Therefore, test results are comparable only if these factors are kept constant. Where water is the test liquid, distilled water is preferable forprecise work, but many tap waters suffice for routine control. Alkalies and acids attack thesize film and destroy it or alter its characteristics. Solutions that are used repeatedly mayleach substances from the paper samples and undergo changes. Where accuracy is required, fresh liquid should be used for each test.

To avoid anomalous data, the physical and chemical characteristics of the test fluid must remainconstant during the test period. For example, if a surfactant is used to increase the rateat which water will penetrate a given web, the surfactant could be sorbed by the fibers and thus the surface tension of the test liquid could be increasing with time. Similarly, the surfactant could be dissolving components of the sheet and in turn this, too, could have a marked effect on penetration. If there is no linear zone such as is shown in B of Figure 7.1, then there is a good probability that there may be a problem relating to the constancy of the test fluid.

Preparation of sample

Paper samples for testing should be handled carefully to avoid creasing or soiling. Rough handling may develop mechanical weaknesses and soiling may change the contact angle. Perspiration or sebaceous oils from the skin may affect sizing as shown by Hudson (46). The importance of conditioning under standard temperature and humidity has already been mentioned. Wire and felt sides of paper may show differences in the rate of penetration. It is customary to test each side of the paper using separate specimens, and average the results.

In a series of studies (50), Hudson's studies were confirmed. To decrease the R^2 value to a minimum in preparing TAPPI handsheets, it was found necessary to wash the plates with alcohol and to handle them only at the edges.

Factors influencing the choice of test method

There appears to be no probability of the development of a single test method that will be best for all cases. It is certainly logical, therefore, to select the method that most nearly indicates the serviceability of the product.

Tests to measure sizing have but two uses: to maximize production on a paper machine, and to give the customer a product that will meet his needs.

Specifically, in the manufacture of commodity grades of paper (bond, offset, mimeo, etc.), the prime need for sizing is to enable the web to pass through the size press of the paper machine without picking up an excess of the size press formulation and without breaking.

In the manufacture of paper and board for use in packaging dairy products, resistance to edge wicking (by the dairy products) is important. Here, it was empirically found that a board that had resistance to the penetration of a 20% lactic acid solution would be resistant to edge wicking by dairy products.

Therefore, today board destined for packaging dairy products must meet a given resistance to a 20% lactic acid solution. The fact that lactic acid is not present to any extent in edible dairy products does not lessen the need to meet that test.

Referee tests applied to determine whether a product meets specification or to check control tests generally are not limited by time considerations and may be devised with greater accuracy and reliability in mind. The same can be said of tests used in research and development work.

Many have wished for a test that would measure the rate of water transudation under little or no pressure and at room temperature. For lightly sized paper or board the dry indicator method serves well because water in liquid form actually passes through the sheet in a short time. On the other hand, hard-sized heavy papers and boards when tested with one side exposed to circulating air resist the transudation of water so long that this method of test becomes impracticable.

Wink and Van den Akker (12) have investigated this evaporation effect with various sheathing boards of 3/4-in. thickness. After exposure of one side to water for various lengths of time, the board was split into four approximately equal sections by sawing in a plane parallel to the surface of the board. One sample was in contact with water at 38°C on one side (the edges having been sealed with wax) for 310 hr (about 13 days). Some of the sectioned layers contained half as much moisture as the board would have held if brought into equilibrium with atmosphere of 100% RH at 38°C. Several other samples that were in contact with water for periods ranging from 23 to 192 hr showed very little penetration of the water in liquid form.

According to Swanson and Van den Akker (13), the mechanism of the penetration of water through a hard-sized sheet is probably as follows:

"When a sheet is placed in contact with water, and the free surface is left in contact with the ambient air (specimen not covered), water slowly penetrates the paper and the rate of penetration slows down, roughly in inverse proportion to the distance penetrated. When penetration through a certain fraction of the sheet occurs, the rate at which water evaporates and diffuses through the residual fraction of the sheet becomes equal to the rate of liquid phase penetration under capillary forces. Thus at steady-state the boundary between water-soaked and incompletely permeated paper is static. The side of the paper facing the air (and the observer) does not assume the appearance and physical properties of water-soaked paper.

"In order that reasonable transudation times for hard-sized papers could be observed with the Fluorescence Test, the decision was made, years ago, to run the test with the specimen covered with a suitable glass plate (transparent to the near-ultraviolet). This almost completely inhibits evaporation, and permits water transudation to be observed (in the case of very hard-sized papers, the time may be 20 min., or longer, as contrasted with perhaps 60 sec. in the TAPPI dry-indicator test)."

The alternative, of course, is to introduce accelerating factors such as heat or chemicals to reduce contact angle. In so doing care must be exercised to make sure that the test so designed gives the proper information. Before final adoption of any method, a check comparing the results it gives with the serviceability of the paper or board is advisable. More detail on this is given in the next section.

Types of tests

Probably no system of classification of size tests is completely applicable to the many tests for water resistance that have been devised. One classification would be to divide the tests into what might be called direct and indirect measurement.

Direct methods involve the actual observance of penetration of water into or through the sheet. In some cases penetration may be determined by weighing the sample, rather than by visual perception.

Indirect methods take advantage of changes in sheet properties as a result of entry of water into the sheet structure. Degree or rate of penetration is determined by the measurement of these properties. An example of an indirect method is the use of the change in the electrical conductance of a sheet with penetration to indicate sizing.

Another way of classifying the available tests would be as to whether the end point is determined by a fixed time or by the degree of penetration. The Cobb Test is an example of the former and the Ink Flotation Test illustrates the latter type.

A third classification would be to divide the tests into three categories: those that measure the penetration of fluids into paper and board; those that measure the resistance of the surface layers of the board to sorption and extrapolate the values obtained to sizing; and, finally, those that measure a secondary property of paper and board and again extrapolate this value to sizing.

Examples of those tests that measure penetration would be the Hercules HST System (27), the Valley Size Test (38), and Ink Flotation (34).

The Cobb Test (15) would be a good example of the second type of test and test procedures, such as the Carson Curl Test (16,17) and Contact Angle (22,23), would be good examples of those tests that measure a property allied to sizing and from which the data is extrapolated to a sizing value.

However, even this falls short when trying to classify a test such as the "Bristow Test" (51,52). The "Bristow Test" (51,52) measures a combination of surface resistance and penetration.

For this paper the Bristow Test and the Vanceometer Test (this is mainly used for measuring the resistance to penetration of oils, but it can be used with aqueous test fluids) are considered sub-categories of the measurement of surface characteristics of papers.

Test procedures

It is neither practical nor necessary to give detailed descriptions of all of the test methods that have received acceptance in the paper industry, particularly those that are described in detail elsewhere. Before setting up tests for sizing, account should be taken of TAPPI Method T 400, "Sampling Paper for Testing," and TAPPI Method T 402, "Conditioning Paper and Paperboard for Testing." Brief descriptions of the more commonly accepted tests are given below with references to more detailed discussions in the literature.

Test procedures

- Those that measure the penetration of aqueous fluids into paper and board

  - Those that measure penetration down through paper and board

●● Those that measure penetration using changes in conductivity

The Currier Size Test

The Currier Test (19-21) depends upon the electrical conductivity of wet paper versus that of dry paper. The sample is placed on a metal plate and covered by a felt disk above which is an inverted bottle of distilled water. The metal plate and felt are connected to a battery which delivers 1 mA of current. Degree of sizing is measured in terms of the time required for the current to reach a given amperage. For papers that contain conductive materials, it is recommended that Japanese tissue be inserted between the sample and the plate. This provides insulation until transudation of water occurs, but offers practically no resistance to water penetration.

This method has the advantage of instrumental measurement of the end point rather than subjective judgement. On the other hand, it is open to the criticism that electrical conductivity is being measured which may not be dependent entirely on water transudation. Furthermore, the effect of the transfer of moisture in vapor form could well be a factor. This possibility is suggested by the fact that Currier size times are normally considerably less than those obtained by the dry indicator test. The procedure for conducting the test is described in TAPPI Useful Method 433.

●●● KBB Test

The KBB Test method and apparatus has been developed in Great Britain and is also referred to as the Galvanic Size Tester. The instrument consists of two disk-shaped electrodes, one of zinc and the other of sintered porous bronze (Porosint). A galvanic cell is formed when two terminals contact opposite sides of a moist sample of paper sandwiched between them. The electrodes are connected to the terminals of a microammeter by which the flow of current is measured. Between uses the bronze electrode is kept lying on a pile of blotting papers which are kept saturated with electrolyte and ready for use at all times.

In conducting the test the sample is placed on the zinc plate and the porous bronze disk is laid on top of the sample. In a short time a small amount of current begins to flow and the size time is taken when the current reaches a certain value. A current of 80 μA has tentatively been chosen as the most satisfactory end point.

The method of test is simple and the end point is sharp. This eliminates the personal element which is present in tests involving the judgement of end points by color development.

Van den Akker and Hardacker (28) have evaluated this method in light of both advocacy (Bridge, Harrison, and Wright (29,30)), and criticism (Harrison, Banks, and Poulter (31)). In general they consider the method to have merit but they point out certain disadvantages. One of these is the tendency of some papers to cockle, buckle, or "saucer" under the influence of moisture. It is suggested that this difficulty might be minimized by increasing the weight of the bronze electrode and/or raising the end point to a time when the sheet contains more water and is therefore more plastic.

The advocates of the method believe that moisture in vapor form plays only a minor part in reaching the end point. On the other hand, time-current curves are frequently nearly straight lines and critics of the method point out that there is no break to show that liquid water has reached the zinc plate. As a rebuttal to this, it is pointed out that water penetrates unevenly and that under such circumstances a gradual increase the current flow is to be expected.

●●● Valley Size Test

This test, originally suggested by Okell (38), is based on the principle that water resistance can be measured by determining the conductivity of paper exposed on both sides to a solution of an electrolyte. It is assumed that as the electrolytic solution penetrates the paper, the resistance to the flow of electricity decreases. This method studied by Carson (39) and Alexander (40) and discounted by both has nevertheless gained acceptance in many mills.

The method consists in placing the sample of paper between two electrolytic cells between which there is a difference of electrical potential, and observing the drop in electrical resistance as the solution penetrates the paper from both sides. The recommended solution consists of normal potassium chloride with some glycerine to reduce the rate of penetration. Sodium chloride may be used alternatively. The apparatus consists of two hard rubber pockets which fit together to make a water-tight chamber. The sample to be tested is clamped between the two pockets. A storage reservoir for the test solution is connected by tube to each pocket and an ammeter to measure the flow of current is provided. In the early stages of development, ammeter readings were taken at intervals and plotted in the form of a curve which revealed the rate of drop of electrical resistance as the electrolyte penetrated the sheet. More recently the test has been simplified so that an end point is read only when the current reaches 160 mA.

Carson (39) points out that the air content of the sheet is a factor in the Valley Test, and Casey (41) confirms this by stating that as the density of the sheet increases, the Valley Test becomes lower. Carson believes that the capacity of the test solution to dissolve air causes effects which are not related to sizing degree. The structure of the sheet also has a bearing on the Valley Test results. Verhoeff and coworkers (5) found that the Valley Test was decreased by beating. This was attributed to an increase in the number of interfiber bonds and intrafiber penetration. As an extreme case the Valley Size Test on unsized cellophane was less than 2 sec. whereas most sized paper handsheet samples showed tests upwards of 20 sec.

The Institute of Paper Chemistry (42) made a careful study comparing the Valley, Currier, and Dry Indicator Tests. It was found that the Currier and Dry Indicator Tests agreed with each other in a general way, but Valley Tests were much more erratic. It was concluded that the Valley Test is not a reliable method of testing paper for water resistance.

- Those that measure penetration using reflectance
  - ●●● HST (Liquid Penetration Tester)
    An apparatus also known as the Hercules Photometric Ink Penetration Tester is described in Price, Osborn, and Davis (27). It measures the penetration of liquids into or through paper by measuring the effect of the liquid on the light transmittance or reflectance of the sample. The sample is placed between two spheres, one of which is suspended horizontally above the other. The upper sphere has an opening in its bottom and the lower sphere an opening in its top. Light from a tungsten lamp is directed through condensing lenses and a filter into the lower sphere from which it is reflected onto the paper sample. The filter provides monochromatic light to eliminate as far as possible the effect of the color of the sample. Photocells in each sphere measure the reflectance and transmittance of the light. These cells are balanced against a photocell that is directly exposed to the lamp through a filter.

    In conducting tests, the test liquid is poured on the sample through the upper cylinder and a stopwatch started simultaneously. As the liquid penetrates the sample, the light transmitted to the upper photocell increases and the light reflected to the lower cell diminishes. Either or both light values are measured and plotted against time. For production control tests where available time is limited, the end point can be taken when reflectance reaches a predetermined value.

    Various liquids may be used. However, as the test procedure is based on the change in the reflectance of the system with time, the fluids must be colored. In general, Calcocid® green dye has been found most satisfactory for normal usage. Studies have shown that anomalous data will develop if the fibers in the paper or board being tested are cationic (the Calcocid® green dye is anionic).

    As shown by the work of Gess (49), (Figure 7.4), the mechanism by which the common accelerators increase the rate at which penetration occurs is not clear. It is therefore strongly recommended that the test fluid used have a meaning over and above

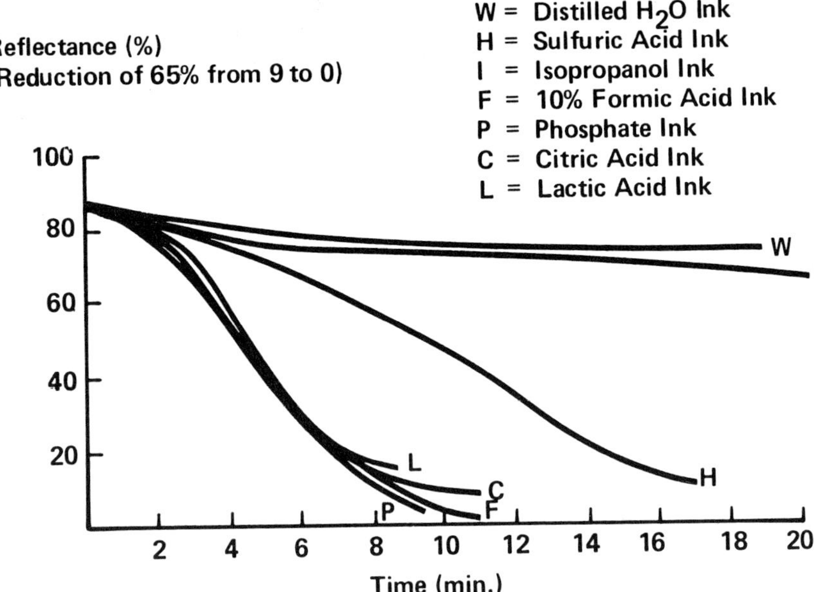

Figure 7.4 Effect of common accelerators on the rate of penetration of a Calcocid® green water solution into sized paper.

that as an accelerator.

Specifically, the standard HST Ink #2 uses formic acid as an accelerator. Formic acid has an effect on the penetration rate other than that of acidity. In addition, it is difficult to understand the use of the acid HST Ink #2 in measuring the sizing of paper containing calcium carbonate pigment and manufactured at pHs in the range of 7.5 plus. The character of the ink will be changing as it penetrates due to the buffering action of the calcium carbonate (reacting with the ink to give calcium ions and carbon dioxide).

The observation is made that a well-sized paper will not achieve complete saturation of the pores with water alone. After soaking a sample in water for four days, transmittance reached only 35% of that with no sample in the instrument whereas a waterleaf sheet made from the same pulp reached 69% transmittance. For well-sized papers accelerative factors are suggested.

●●● Other test methods using reflectance

Since the work of Osborn, etc. (27), many other units have been developed to measure the change in reflectance of a paper or board as a dyed fluid penetrates. These would include, among others, the Photosize Liquid Penetration Tester, the Penetrace Liquid Penetration Tester, and the Pen- A-Size Tester.

It should be noted that all the units that measure penetration have one major weakness: they must be able to "see" the top surface of the board or paper being tested to be able to measure the initial penetration rate of the test fluid. If the board is too thick, then the dye will have to penetrate a given distance before the unit can actually begin measuring true penetration (i.e., it can begin to "see" the dye front).

When penetration of a thick board is measured with a recorder, the initial penetration will be seen as a flat line (i.e., it will look as if penetration is not occurring).

●● Those that measure penetration up through paper and board

  ●● Water Flotation Test
Codwise (10) has studied size testing by flotation of the sample on water, the temperature of the water being varied from 72 to 212°F. Apparently no indicator was used to facilitate the observation of the appearance of water on the upper face of the sample. In the case of newsboards, the darkening of the surface by the transuded water provides a sharp end point but it should be possible to use dry indicator if required. Codwise's work has already been discussed under the heading "Effect of Temperature."

  ●●● Dry Indicator Test
In an effort to develop a test which would measure only the transudation of water through a sheet of paper, Carson (18) devised the dry indicator method. The method depends on the fact that when water-soluble dye powders are mixed with powdered sugar and moistened, a sharp development of color takes place. When one side of the sample is exposed to liquid water and the indicator dusted on the other side, a development of color in the indicator means that water has passed through the sheet and the process can be timed to give a numerical value. The test is described in detail in TAPPI Method T 433 and in modified form as TAPPI Useful Method 421.

The dry indicator method has the virtue that the penetration of water only is measured eliminating effects of dissolved salts or dyes used in some other test methods. The end point is determined by the actual observation of the exudation of water rather than noting the effect of the penetration of the water on some other property of the paper such as strength or opacity. This method has received more study by TAPPI members and committees than any other and in the past has been widely used for papers where transudation times are not greater than a very few minutes.

For very hard-sized papers and boards the dry indicator method is open to serious objection. First, the end point develops slowly and is very subjective. This renders it difficult to secure agreement among various observers.

The effect of moisture vapor is more serious. Van den Akker, Nolan, Dreshfield, and Heller (9) have shown that when the dry indicator is exposed to water vapor only, the development of color takes place in time as if it had contacted liquid water. It has been demonstrated that with hard-sized papers the dry indicator end point is reached well before liquid water has gone through the sheet. The influence of water vapor increases rapidly with degree of sizing above half a minute's transudation time. Since internal sizing and water vapor resistance are achieved by entirely different principles, it is highly important to the paper manufacturer to measure the two properties separately.

●●● Fluorescence Size Test

In order to eliminate the influence of moisture vapor, which affects the dry indicator test, Van den Akker, Nolan, Dreshfield, and Heller (9) developed the Fluorescence Size Test using a nonhygroscopic indicator. The test is based on the principle that certain dyes such as Rhodamine 6 GX and Uranine B fluoresce under ultraviolet light when in a solution of sufficiently low concentration. The concentration of the dye solution at which fluorescence occurs is important to prevent the development of an end point as a result of the absorption of water by the surface fibers of the paper from water-vapor-laden air. In other words, liquid water must come through the sheet in sufficient amount to dissolve the dye to sufficiently low concentration to produce fluorescence under ultraviolet light.

The test was originally carried out by applying the dye powder to the sample in a very small amount with a camel-hair brush. The sample was then floated on water inside a closed box to which ultraviolet light was admitted through an aperture in the side of the box. Observation of the progress of the test was made through another aperture in the top of the box. Uranine B dye was used and the end point was taken as the time when half the treated area fluoresced a uniform bright yellow.

Because of the subjective nature of the readings and the difficulty of securing agreement among various observers a mechanism was designed to avoid the personal factor. Observation of the progress of the development of fluorescence is made by means of an eyepiece of which half the field is illuminated by light reflected from the sample. The other half of the field is illuminated by light from a reference source which has passed through a series of polaroid disks and reflected into the eyepiece. The two halves are matched by rotating one of the polaroid disks. The degree of rotation of the disk, plotted against time, gives a curve from which a numerical result may be derived. The Fluorescence Size Tester is now manufactured by the Martin Sweets Co.

●●● Capillary Rise

The capillary rise or Klemm Test (25) is not frequently used for sized papers but has application for absorptive products. It consists of immersing the end of a vertically positioned strip of paper or board in liquid and noting either the time for the liquid to rise to a certain height or, more frequently, the degree of rise in a given time. The test differs from most in that penetration parallel to the plane of the sheet is measured, rather than transudation perpendicular to the sheet surface. Care must be taken to cut the strip in a definite direction with respect to the grain of the paper because the alignment of the fibers has a bearing on the test. Simmonds (26) states that testing of calender-sized paper with water by this method is too slow.

Analysis of the results of this test may be difficult, especially if the front and back sides of the sheet are "sealed." If the surfaces being sealed are not extremely smooth, there will be air pockets between the sealing surface and the surface of the board being tested.

When this occurs, there will be a tendency for the test fluid to rise via the air pockets between the test and sealing surface and this will give fallacious results.

Therefore, when testing boards having sealed surfaces, it is advisable to determine where the advancing front is by cutting into the test board rather than taking the surface measurement.

- Barss Knobel & Young Tester

An instrument familiarly known as the **BKY** News Penetration Tester (36) is designed to provide instrumental measurements of the rate of penetration of colored liquids through paper. It may be considered a device for mechanizing the writing ink flotation test although liquids other than writing ink may be used, such as aqueous dye solutions and dye-colored oils. The instrument is based on the change in reflectance of a sample of paper due to the penetration of a dark liquid coming through from the side opposite that which is being observed. An electric light is so placed that light is reflected from the paper sample into one photocell and from a calibrating surface onto another. The instrument is first "calibrated" with the sample placed over a black surface and observation is made of the *change* in reflectance after the sample is in contact with the test liquid rather than an absolute end point degree of reflectance. A galvanometer measures the difference between two photocell currents. At the start of the tests the currents are not in balance and the galvanometer is deflected. As the surface of the sample darkens due to the penetration of the dyed liquid, the currents are brought into balance and the galvanometer reads zero. This is taken as the end point. The end point may be varied by inserting a black wire screen between the lamp and one of the photocells.

The Institute of Paper Chemistry (37) has evaluated this instrument and has noted that readings are dependent on the color and opacity of the sample being tested. This is considered a serious objection if dissimilar papers are being compared. Of course, if the dye used is absorbed at the surface of the sample as the liquid enters, the situation is similar to that of ink flotation tests where dye absorption capacity of the sample affects results.

● ● ● Penescope

The Penescope Test devised by Allen Abrams (32,33) is not best described as a test method, but rather as an apparatus that can be used for the application of a number of methods. It consists of a cylindrical metal chamber one end of which is threaded to receive a hollow screw cap to clamp the sample in place. A circular sample of paper or board is cut 2.5 in. in diameter and clamped by the screw cap to form one wall of the chamber. The chamber can then be filled with any test liquid and the time noted for the liquid to penetrate the sample.

Penescope tests can be conducted in various ways. The apparatus may be mounted on a panel and the test liquid introduced through tubes or manually through a funnel. It can be used unmounted. In this case, it is partially filled with test liquid, the sample put in place, and the Penescope inverted to initiate the test. It may be used to measure the time for transudation or degree of partial penetration of the test liquid. In the latter case the instrument is mounted on a panel and a graduated tube is connected to the upper side of the chamber. After clamping the sample, the chamber is filled with test liquid to the zero point on the graduated tube. Sizing is then measured by the amount of test liquid absorbed in a given time as measured by the drop in liquid level in the graduated tube. If desired, a curve may be drawn of volume absorption versus time.

Temperature of the test liquid can be controlled by using a jacketed instrument with automatic control of the temperature of water circulating through the jacket (7,8).

The sample may be exposed to ambient air or a glass disk may be clamped over the specimen to inhibit evaporation. If partial penetration is being measured as described above and the surface of the sample is not covered by a glass plate, allowance must be made for bulging of the sample due to hydrostatic head. This has been shown to be a small factor.

● ● ● Lactic Acid Test

For hard-sized boards such as are used in food containers, it is common practice to test sizing in the Penescope using a lactic acid solution as the penetrating medium. In fact, Abrams suggested the use of lactic acid when first describing the Penescope (32) as a method for determining the resistance of paperboard to cottage cheese.

In carrying out the test, a circular sample of 2.5 in. in diameter is cut and one face is painted lightly with 0.1% Methyl Orange. When the sample is placed in the Penescope, the side coated with

Methyl Orange is on the outside. The Penescope is then filled with 140 ml of 20% lactic acid at 20°C. The time of penetration is taken as that between the first contact of lactic acid with the sample and the appearance of six red spots on the outer surface due to the action of the lactic acid on the Methyl Orange.

Methyl Orange is a surface active agent and when the lactic acid first appears as pinpoints, the Methyl Orange causes the acid to spread on the surface. This is avoided by eliminating the Methyl Orange and using Methylene Blue dissolved in the lactic acid solution which appears on the unexposed surface as the lactic acid solution transudes. One gram of Methylene Blue per gallon of solution is the recommended concentration.

The above conditions were established after a study of variations of temperature and concentration of lactic acid. Libby and Casciani (7) also studied these variables. Reference has already been made to their findings regarding temperature effects. They confirmed Abrams' choice of 20% as the best concentration of lactic acid. Lower concentrations prolonged the test excessively and higher concentrations of the acid penetrated too rapidly. However, they suggested the possible use of higher concentrations for heavy, hard-sized boards.

A number of years ago the men who supervise paper testing at American Cyanamid's Stamford Laboratories discovered that as 20% lactic acid solutions age, penetration times increase. Investigation revealed that this was due to the presence of lactides in the lactic acid as purchased. Lactides tend to increase the rate of penetration of the solution. On long standing in 20% solutions, they gradually hydrolize to lactic acid and the rate of transudation falls. In order to eliminate this variable, a method has been devised to hydrolyze the lactides to lactic acid when the solution is first prepared.* This is accomplished by heating an aqueous solution of the lactic acid for an extended time.

Reagents
    Lactic acid, 85% to 90% reagent grade
    Hydrochloric acid, 0.5 N standardized solution
    Sodium hydroxide, 0.5 N standardized solution
    Bromocresol purple indicator -
        dissolve 0.12 g indicator in 100 ml ofalcohol
    Copper sulfate, $CuSO_4 \cdot SH_2O$

---

* Private communication from American Cyanamid Company Laboratories, Stamford, Conn.

Procedure.** Dissolve 2 lb of lactic acid in 2700 ml of distilled water in a 4- liter Erlenmeyer flask. Cover with an inverted beaker and heat on a steam bath for 48 hr. Cool and mix.

Pipette 2 ml of the solution*** into a 250 ml Erlenmeyer flask, add 100 ml of distilled water and 0.5 ml of bromocresol purple indicator. Titrate with 0.5 N NaOH to a full purple color. Calculate the dilution necessary to give a titration of 9.3 ml, add this amount of water, and mix. Repeat the titration and record the volume used (volume A); this value should be between 9.1 and 9.6 ml.

Calculation:

$$\% \text{ lactic acid (by weight)} = \frac{A \times N \times 9.0}{2 \times 1.05}$$

where

A = volume of NaOH used for titration of 2 ml of solution to the first end point

N = normality of NaOH

To the neutral solution obtained in the final titration above, add 3.00 ml of 0.5 N NaOH and boil for 3 min. Cool and add 3.00 ml of 0.5 N HCl. Titrate with 0.5 N NaOH to the original full purple color; the volume of 0.5 N NaOH required for the last titration should be no greater than 0.1 ml. Add 6 g of $CuSO_4$ $\cdot H_2O$ to prevent mold growth.

●●● Writing Ink Penetration (Ink Flotation) Test

One of the most commonly used tests for size resistance is ink flotation. Its popularity is probably due to its seeming simplicity. A number of variations of the test are possible but at its best it consists in cutting a sample about 2 in. square, folding the edges upward to avoid penetration of the ink through the edges, floating the sample on standard writing ink at a controlled temperature, and noting the time for a specified degree of color to appear on the upper surface. The test is easily set up and carried out but there are a number of objections to it.

A well-known criticism of the test is the indefiniteness of the end point which makes it difficult to secure close agreement between different operators. More serious is the fact that the penetration of the water in the ink through a highly sized paper is more rapid

---

** The procedure given is for the preparation of about one gallon of solution. Larger volumes should be prepared in 1-gal. batches following the same initial treatment; after heating on a steam bath, cool and blend in a suitable container before testing.

*** For greater accuracy the 2 ml of solution should be weighed and this value used in the denominator of the calculation. The final concentration should be between 19.5% and 20.5% by weight.

than the progress of the color. Therefore, in part at least, what is being measured is the capacity of the paper to absorb the dye in the ink.

It may be shown that dye absorption is promoted by rosin sizing entirely apart from the water resistance of the paper.

Wilson (34) studied the mechanism of the penetration of writing inks and suggested that the aluminum compound in the paper coagulates the dye in the ink. His work corroborates the fact that ink penetration tests do not correlate with water resistance properties but do provide information concerning the behavior of paper toward writing inks.

- Those that measure surface properties of paper and board and extrapolate the results to sizing

  - Cobb Test
    This test was devised by Miss R. M. K. Cobb and first described by Cobb and Lowe (15). It consists of exposing one surface of the sample to water for a given time and measuring the amount of water absorbed by the gain in weight. The water is confined to a fixed area of the sample (usually 100 cm^2) by means of a metal ring clamped to the sheet. Details of the apparatus and method are given in TAPPI Method T 441.

    While the approved method calls for reporting results in terms of weight of water absorbed per square meter, Cobb and Lowe describe a method of determining the average depth of penetration. The method is designed to show surface resistance, but by testing both sides of the sheet a very good idea of the total size resistance is obtained. Cobb and Lowe give the following advantages of the test.

1. An end point in a reasonable length of time is a certainty.
2. Results are readily duplicable by different operators.
3. A fairly large area is tested, which minimizes the effect of local variations.
4. The method is flexible. Practically any test liquid may be used and temperature variations are possible. Also the time may be varied to suit the type of paper or board being tested.

The disadvantages are:

1. Constant attention must be given the test, but since the time is usually short, this criticism is not serious.
2. Manipulation requires care and reasonably good technique.
3. The test is not readily applicable to thin or absorbent papers.

The Cobb Test is at its best when measuring the so-called medium-sized papers (those having 1 min. Cobb values between 28-65 gms/m^2 of water sorption).

With hard-sized papers, the 1 min. Cobb tends to be a measure of the surface area of the board or paper being tested (i.e., the smoothness of the board).

At the present time, the Cobb Test (10 min. Cobb) is about the only means of measuring the sizing of unbleached boards.

Contact Angle Test

Cobb and Lowe (15) point out that only sizing materials that raise the contact angle between a drop of water and the paper sample are effective against aqueous liquids. It is therefore informative to measure the contact angle as a means of determining the surface properties of paper. Lafontaine (22) has devised a method of making the measurement with facility and precision and this has been evaluated by Codwise (23). The method has since been adopted as TAPPI Method T 458.

The angle measured is that made by a drop of water, ink, or other liquid resting on a paper or board surface 5 sec. after deposition of the drop. The angle reported is the one on the drop side of a tangent drawn to the surface of the drop at the point of contact with the paper. Thus, the contact angle becomes larger as the drop approaches a spherical form (see Figure 7.5).

It might be thought that the contact angle would be a direct measure of the chemical and physical nature of the sizing materials in the surface of the paper regardless of the mechanical structure of the sheet. However, Lafontaine finds that the degree of calendering has a definite influence on the contact angle; the higher the finish the lower the angle. Calendering would not be expected to change the chemical nature of the sizing, but it could well alter the amount of unsized fiber surface exposed and affect the amount of air trapped under the drop.

Codwise (23) found a fair correlation between the angle of contact, the drop test (drop spreading), and the Cobb Test (on papers not tub sized).

Foote (24) devised a technique for determining the contact angle of a single fiber partially immersed in water in conjunction with a study of average pore diameter. He found a relationship between contact angle and the TAPPI dry indicator method.

Tests that measure a combination of surface and absorption

Drop Test

This is a test for surface water resistance but under some circumstances the interior of the sheet may influence the test. The test consists of placing a small but definite quantity of water or other liquid on the surface of the sample

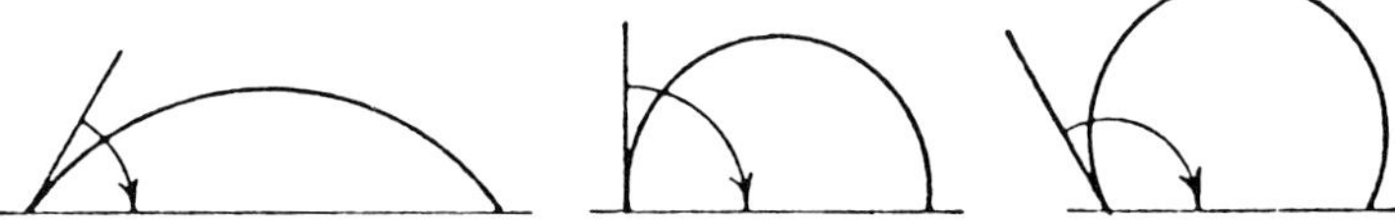

Figure 7.5 Contact angles on paper surfaces. (a) Acute angle. (b) 90° angle. (c) Obtuse angle

in a prescribed manner and timing either the beginning of the spreading of the drop or the total absorption. If the time for completion of the test is more than a very few minutes, it is recommended that the drop be covered with a watch crystal to prevent evaporation of the water. The test is especially adaptable to moderate- or slack-sized boards and is also used for testing of bibulous papers.

Many variations of the drop test are possible. In any case the test should be conducted under conditions of controlled temperature and humidity. Codwise (14) suggests the use of 1 ml of water delivered from a pipette, the tip of which is held close to the sample so that it is immersed in the water during delivery. The pipette is then removed and if the sample is moderately or well sized, the water is covered with a watch crystal. Usually the end point chosen is the first appearance of the spreading of the water in the surface of the sample. For very hard-sized samples the test may run 24 hours or more. For slack-sized papers or boards, it may be desirable to run the test until the drop is completely absorbed as evidenced by the disappearance of the glossy surface of the water.

The drop test is simple and requires little equipment. Three procedures for conducting the test are described in TAPPI Method T 432, TAPPI Suggested Method T 492. It is particularly useful in testing cylinder boards where the main consideration is the sizing of the liner. On the other hand, because of the duration of the test it is not a practical means of distinguishing between degrees of sizing in the very high range.

The Bristow Tester

The unit (Figure 7.6) is specifically designed to measure the sorption of water by essentially waterleaf sheets.

Figure 7.6A

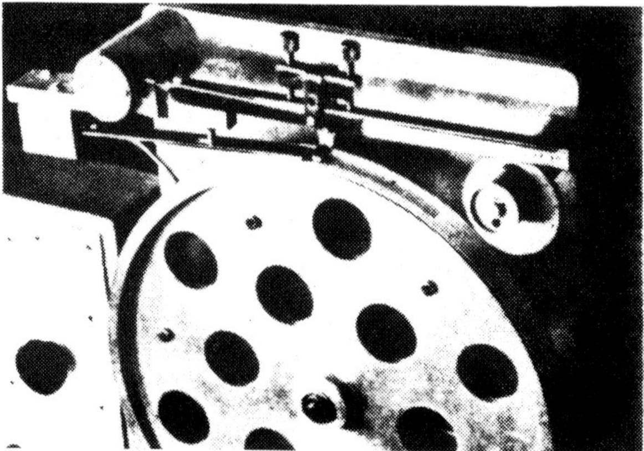

Apparatus for the study of liquid absorption during short time intervals. A paper sample is fixed onto the circumference of the wheel and a known quantity of liquid is spread onto the paper from the holder directly above the wheel. The variation in length of the track at different speeds enables the absorption coefficient and roughness index to be determined.

Time Available for Sorption $= \dfrac{L}{V}$

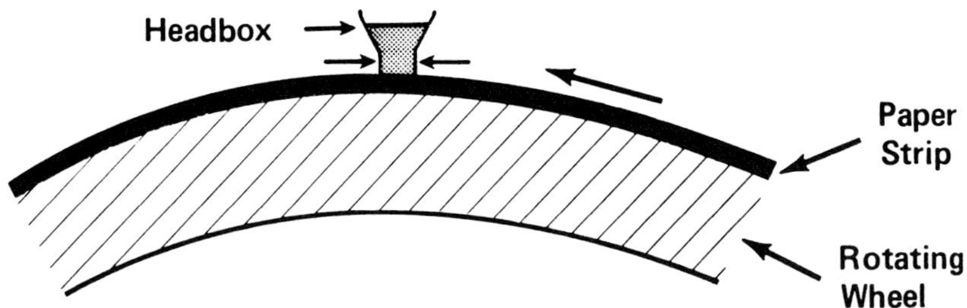

**Schematic diagram of Bristow's device for measuring wetting
and sorption. Headbox is filled with a known amount of liquid
and rests on the test paper strip with a pressure of 0.1 MPa.
The wheel can rotate at several precisely controlled speeds V.
[TAPPI 65, 98 (1982)]**

Figure 7.6B The Bristow test

In the operation of the unit, a metered amount of fluid is transferred to
a sheet of paper mounted on a wheel rotating at a constant speed.

The paper being tested is analyzed by the equation:

$$^{V}/_{LB} = Kr + Ka \; \pi^{\frac{1}{2}}$$

where

V	=	the volume of liquid used (ml)
L	=	the length of the track produced (m)
B	=	the length of the slot (m)
π	=	the absorption time (s)
Kr	=	the roughness index (ml/m^2)
Ka	=	the absorption coefficient (ml/m$^2 \cdot$ s$^{\frac{1}{2}}$)

The time π is given by the expression D/S where D is the tangential width
of the slot and S is the speed at the circumference of the wheel. By analyzing
a series of runs (varying quantities of liquid and varying RPMs on the wheel),
it is possible to arrive at Kr + Ka. Ka is essentially the "sizing" value of the
sheet being tested.

Kr has been found by Bristow to correlate well with Bendtsen Roughness.
Bristow's unit was modified by Lyne and Aspler (52) and is being used to
measure the "sizing" of newsprint and of the base paper made for the
manufacture of LWC (light weight coated) grades of paper.

The Vanceometer Test

The Vanceometer was originally developed to measure the sorption of the oils used in the manufacture of nonaqueous printing inks. It, however, can be used in the manner of the Bristow Tester to qualitatively measure the sorption of water by various grades of paper.

To this variation of the Vanceometer Test, a measured amount of fluid is placed on the paper and the wheel released. By measuring the area of the spot formed, it is possible to have a good measure of the absorptivity of that board.

This type of rapid test is useful in mills where it is important to have data to react rapidly and prevent the loss of paper due to something in the manufacturing process. It should be emphasized that this is not the type of test that will give the quantitative data required in research, but it is a good in-mill test.

Those that measure properties related to the sorption of water and extrapolate the data to sizing

- Carson Curl Test

  This method of testing the sizing of paper was developed by F. T. Carson (16,17) of the U.S. Bureau of Standards. It is based on the fact that when paper is floated on water the edges curl upward, the axis of the curl paralleling the machine direction of the paper. After an interval which is related to the degree of sizing, the edges of the sample reverse the direction of movement and the sheet tends to flatten.

  Carson proposed that the maximum curl is reached when the water has penetrated halfway through the sheet on the average and the curl reverses when the water passes the halfway point. Therefore, by timing the interval between the first contact of the sample with the water and the point of maximum curl a measure of water penetration is obtained. The curl method is not well adapted to the testing of laboratory handsheets since they do not exhibit a grain pattern and the curl is not so pronounced as with machine-made papers. The method is described in detail as TAPPI Method T466. A modified procedure is given in TAPPI Useful Method 459.

- Pen and Ink Tests

  When writing paper is being tested for size resistance, the main consideration is to determine how the paper will behave when written on with pen and ink. It is therefore not surprising or illogical that such paper is tested by drawing lines with a pen and noting the degree of spreading of the lines while the ink is drying. The test is extremely useful but lacks precision, and, as usually conducted, gives no numerical figure for the degree of sizing. It is, of course, possible to draw the lines with a draftsman's rulling pen and measure the degree of spreading with a microscope with a graduated scale. With any kind of pen a number of variables enter the picture which are difficult to control precisely such as: pressure applied, sharpness of the pen, angle of the line with the grain of the paper (cross-grain lines spread more than machine- direction lines), and manner of holding the pen. Other variables that are controllable are atmospheric temperature and humidity and composition of the ink.

Schur and Levy (35) have devised an apparatus to provide control of the variables enumerated. It consists of a small wheel with a slot extending around its perimeter in which the ink is placed. The wheel is mounted on a carriage in such a way that the wheel can be drawn across a sample of paper (lying on a glass plate) with a fixed pressure and controlled rate of speed. Spreading is measured with a graduated microscope. TAPPI Useful Method 423 provides a procedure using a pen and a solution of Malachite Green.

It should be pointed out that, as in ink penetration tests, coagulation or absorption of the dye plays a prominent part. The ideal writing paper is one that absorbs the water in the ink but fixes the coloring matter before it can spread. It has been shown that as size and alumina content of the paper are increased, water resistance tends to reach a maximum, but the fixing power for ink colors continues to increase. This is fortunate for one desiring to produce the ideal writing paper.

- Water absorption by total immersion

Heavy boards that are moderately or highly sized are sometimes tested by immersing the sample completely in water. The amount of water absorbed is determined by weighing the sample before and after immersion. The sample is preferably conditioned according to TAPPI Method T 402, but if the test is made for production control and time is an important consideration, the sample should at least be cooled before immersion. Otherwise a vacuum effect may occur due to cooling of the air in the sample after submergence. Water temperature must be closely controlled and distilled water is preferable, but tap water may be used if it is of neutral pH. The sample should be blotted gently with a soft towel after lifting from the bath to remove surface water. The blotting procedure and the time interval between blotting and weighing must be carefully standardized to provide comparable results.

This test is described as TAPPI Suggested Method T 491 and the suggested time of immersion is 10 min. The time can be varied at will to suit the type of product being tested. Insulating board may require several hours in order to provide meaningful data.

An important consideration in the Total Immersion Test is the air entrapped in the pores of the sample (45). If deaerated water is used for bath, penetration is definitely more rapid than if water is used that has an equilibrium content of dissolved air. Obviously the capacity of the water to dissolve the air in the pores of the sample has an important bearing on the rate of penetration. This has also been noted in connection with the Valley Test. Where the water source is always the same, results should always be comparable if time, temperature, depth of submersion, and size of sample are kept uniform. Figure 7.7 shows curves illustrating the effect of air in total immersion tests.

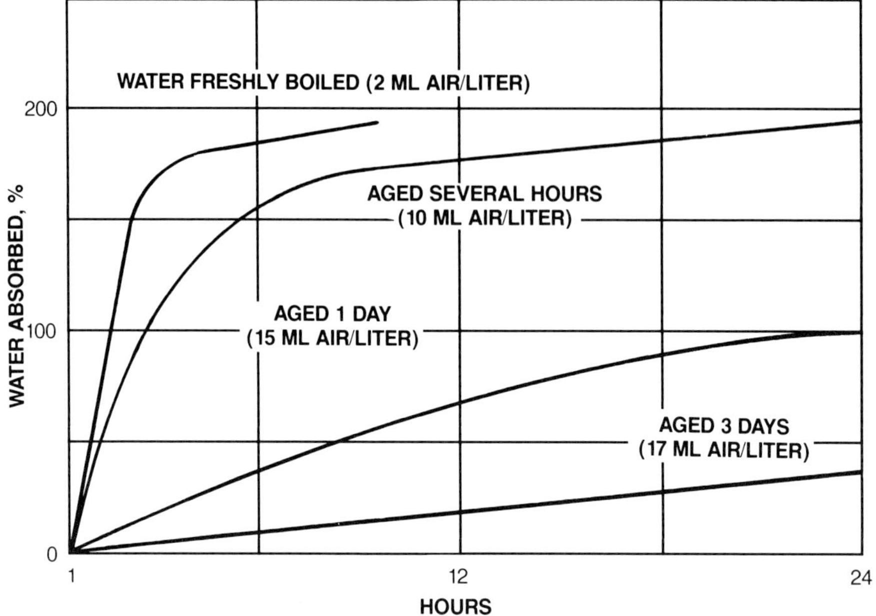

Figure 7.7 The effect of dissolved air on water absorption of paperboard (50% bleached sulfite; 2% size).

- Blood Serum Penetration Test

In order to determine the suitability of various paper products for the packaging of raw meats, a satisfactory test is necessary. Libby and Parkinson (44) have shown that a number of previously established sizing tests (ink flotation, Valley, lactic acid, water flotation) give no useful information concerning the resistance of paper to blood and blood serum. In the past it has been common practice to test this property with the use of raw beef liver. However, it is difficult to control the variables precisely and to secure agreement among independent operators. A test using hog liver is described in TAPPI Suggested Method No. 486.

Libby and Parkinson evaluated the use of blood serum, which may be obtained as a standard article from meat packers. They used the Abrams Penescope in carrying out the test employing bromocresol purple indicator to show the end point. This indicator is yellow in the pH range of sized paper, but is purple at pH 6.6 to which the blood serum was adjusted. The blood serum test agreed closely with parallel tests conducted with liver.

TAPPI Useful Method 446 describes a modification of the method presently by Libby. Sodium carbonate is added to the blood serum which renders the serum decidedly alkaline. Bromothymol blue is then used as the indicator by painting it on one side of the sample and allowing it to dry. The sample is then placed on a glass plate with the indicator next to the glass. A small brass cylinder is then placed on top of the sample and a few drops of the blood serum are placed in the cylinder. The time for a blue color to appear as observed through the glass plate is the measure of resistance to blood serum.

Literature Cited

1. Carson, F.T. 1949. *Tappi.* 32(11):501.
2. Billington, P.S. and Hrubesky, C.E. 1931. *Tech. Assoc. Papers.* 14 (1):273.
3. Baird, P.K. and Hrubesky, C.E. May 1930. *Tech. Assoc. Papers.* 13(1):274.
4. Wilson, W.D. 1949. *Tappi.* 32 (9): 429.
5. Verhoeff, J., Hart, J.A. and Gallay, W. 1963. *Pulp and Paper Mag. of Canada.* 64: T509.
6. Cobb, R.M.K. 1935. *Tech. Assoc. Papers.* 18(1): 290.
7. Libby, C.E. and Casciani, F. 1932. *Tech. Assoc. Papers.* 12 (1): 263.
8. American Cyanamid Company. 1963. "Sizing Technical Highlights." A7.
9. Van den Akker, J.A., Nolan, P., Dreshfield, A.C. and Heller, H.F. 1940. *Tech. Assoc. Papers.* 23(1): 569.
10. Codwise, P.W. 1943. *Tech. Assoc. Papers.* 26(1): 165.
11. Codwise, P.W. 1931. *Tech. Assoc. Papers.* 14(1): 175.
12. Wink, W.A. and Van den Akker, J.A. 1947. *Tech. Assoc. Papers.* 30 (1): 330.
13. Swanson, J.W. and Van den Akker, J.A. Private communication.
14. Codwise, P.W. 1930. *Tech. Assoc. Papers.* 13: 200.
15. Cobb. R.M.K. and Lowe, D.V. 1934. *Tech. Assoc. Papers.* 12 (1): 213.
16. Carson, F.T. 1927. *Paper Trade J.* 79(17): 44.
17. Carson, F.T. 1927. *Paper Ind.* 9(2):259.
18. Carson, F.T. 1925. *Tech. Assoc. Papers.* 8:91.
19. *Paper Mill News.* 1942. 65 (10): 26.
20. Broadbent, F.D., Brown, C. and Harrison, H.A. 1936. *Paper Trade J.* 103 (10):32.
21. Broadbent, F.D., Brown, C. and Harrison, H.A. 1940. *World's Paper Trade Rev.* (Convention Issue): 55, 63.
22. Lafontaine, G.H. 1941. *Paper Trade J.* 113 (6): 29.
23. Codwise, P.W. 1939. *Tech. Assoc. Papers.* 22: 246.
24. Foote, J.E. 1940. *Tech. Assoc. Papers.* 23: 558.
25. Liesegang, R.E. 1943. *Wochbl. Papierfabr.* (6): 219.
26. Simmonds, F.A. 1934. *Tech. Assoc. Papers.* 17: 401.
27. Price, D., Osborn, R.H. and Davis, J.W. 1953. *Tappi.* 36(1): 42.
28. Van den Akker, J.A. and Hardacker, K.W. 1951. *Tappi.* 34 (10): 137.
29. Bridge, F., Harrison, H.A., and Wright, A.V. 1947. *Proc. Tech. Sect.* Papermaker's Assoc. of Great Britain and Ireland. 28 (1): 281.
30. Bridge, F. and Harrison, H.A. 1947. *Proc. Tech. Sect.* Paper Maker's Assoc. of Great Britain and Ireland. 28(2): 661.
31. Harrison, V.G.W., Banks, W.H. and Poulter, S.R.C. 1947. *Proc. Tech. Sect.* Paper Maker's Assoc. of Great Britain and Ireland. 28 (1): 337.
32. Abrams, Allen. 1927. *Paper Trade J.* 84 (3): 44.
33. Tappi Standard T. 441.
34. Wilson, W.S. 1946. *Tech. Assoc. Papers.* 29: 236.
35. Schur, M.O. and Levy, R.M. 1946. *Tech. Assoc. Papers.* 29: 160.
36. Hammond, J. 1936. *Paper Trade J.* 103 (21): 37.

37. Institute of Paper Chemistry. "Instrumentation Studies." No. 24; 1937. *Paper Trade J.* 105 (26): 35.
38. Okell, S.A. 1917. *Paper.* 20 (5): 20,32.
39. Carson, F.T. 1923. *Paper Trade J.* 76 (15): 187.
40. Alexander, J.E. 1923. *Tech. Assoc. PApers.* 6: 37.
41. Casey, J.P. 1960. *Pulp and Paper.* 2 ed. New York: Interscience. Vol. 2, p. 1070.
42. Institute of Paper Chemistry. 1937. "Instrumentation Studies." No. 16; *Paper Trade J.* 104 (23): 45.
43. Stockigt, F. 1920. *Paper.* 26 (1).
44. Libby, C.E. and Parkinson, L. 1933. *Tech. Assoc. Papers.* 16(1): 386.
45. American Cyanamid Co. 1963. "Sizing Technical Highlights - Air, a Factor in Water Resistance Tests." A-11, A-12.
46. Hudson, F.L. and Eynon, D.L. 1965. *Paper Tech.* 6(3): 224.
47. Reaville, E.T. and Hine, W.R., Jr. 1967. *Tappi.* 50(6): 262.
48. Van den Akker, J.A. and Wink, W.A. 1969. *Tappi.* 52 (12): 2406.
49. Gess, J.M. 1981. *Tappi.* 64(1), 35.
50. Gess, J.M. Personal Communication.
51. Bristow, J.A. 1967. *Svensk Paperstidn.* 70, 623.
52. Lyne, M.B. and Aspler, J.S. 1982. *Tappi.* 65, 98.

Appendix
Mechanisms of Paper Wetting

John W. Swanson

Introduction

One of the important use properties of many paper and paperboard products is their resistance to wetting and penetration by liquids such as water, hot and cold drinks, ink, blood, milk, citrus juices, oils and greases and organic solvents. This paper penetration is known as the "degree of sizing" or "size resistance" or simply the "size test."

Two general methods are used for sizing:

1. Internal or engine sizing consisting of the addition of potentially hydrophobic (from the greek, water hating) materials to the papermaking stock before the sheet is formed. Rosin (and alum), waxes, fatty acid derivatives, hydrocarbon resins, and asphalt are used for sizing toward aqueous liquids. Where oils, greases, and organic solvents are of concern, oleophobic (oil hating) materials are needed such as the perfluorocarbons and silicones.

2. Surface sizing, size pressing, tub sizing or calender sizing consists of the application of film forming polymer dispersions to the surfaces of already formed paper or board at the appropriate location at the dry end of the paper machine. Although these polymer dispersions may sometimes contain small amounts of materials normally used for internal sizing, surface sizing is often done for other purposes such as:

   a. Controlling surface porosity for printing
   b. Decreasing surface fuzz
   c. Improving surface finish
   d. Improving surface picking resistance
   e. Increasing certain strength properties such as tensile strength, folding endurance, and erasure resistance.

Papers that are to be surface sized often already contain internal size in order to regulate penetration of surface sizing polymers into the sheet structure.

Two additional sizing methods, vapor sizing and self-sizing, should also be mentioned. Vapor sizing refers to the treatment of paper with the vapor of a chemical which adsorbs and/or chemically reacts with the fiber surfaces to develop sizing. Certain silanes (1) and fatty acid vapors (2) have been suggested. Photosizing of paper or olefin vapors has been examined by Desai and Shields (98) and by McKelvey and Leekley (99).

Self-sizing refers to the gradual development of sizing in a paper product where sizing is not usually desired. This sometimes occurs in absorbent sanitary products and in corrugating medium. Apparently, self-sizing is caused by the vapor phase redistribution of fatty and resin compounds which are naturally present in all wood pulp products (2). Needless to say, self-sizing may complicate normal internal sizing and may cause adhesion problems (3, 4, 5).

This chapter is concerned with the mechanisms of wetting of paper surfaces by liquids and the factors which affect wetting and penetration.

Wetting of solid surfaces by liquids

Attractive molecular forces, surface tension and surface energy

The molecules of all phases of matter (liquids, solids and vapors) attract one another by finite forces characteristic of their chemical composition. These forces, sometimes called Van der Waals' or secondary valence forces (6) are responsible for the condensation of vapors to liquids and solids, for the cohesive strength of condensed phases of matter, the need for correcting the ideal gas law, the Joule-Thompson effect and the wettability of solids by liquids. It is the latter with which we are most concerned here.

Atoms and molecules within bulk phases are surrounded on all sides more or less uniformly by similar atoms or molecules and the attractive forces for one another are generally in balance. However, molecules which exist at the surface of a solid or liquid generally do not exist in a balanced force field. The molecular density of a liquid surface in equilibrium with its vapor is usually greater than the density of the vapor. Thus, the surface molecules are subjected to a resultant inward attraction and the surface behaves as though it were under tension—called the surface tension. Work is required to form additional surface area against the inward attraction of bulk phase molecules. Therefore, the surface molecules possess more free energy—called the surface free energy—than bulk phase molecules.

The work necessary to bring enough molecules to the surface to form one cm^2 of new area is called the specific free surface energy. Generally the units are ergs/cm^2 and the symbol is γ. For pure liquids the specific free surface energy in ergs/cm^2 is numerically equivalent to the surface tension in dynes/cm. (In SI units the surface tension is given in millinewtons per meter, mN m^{-1} and the specific free surface energy in millijoules per square meter, m^J m^{-2}). Of course there are corresponding interfacial energies between other phases of matter such as solid/liquid, solid/gas, liquid/liquid, and solid/liquid/liquid (immiscible).

When a nonvolatile pure liquid drop is placed upon a clean, smooth, insoluble, plane solid surface the different species of molecules interact with one another to establish a balance of forces. If the adhesional forces between liquid and solid are greater than the cohesional forces of the liquid, then the liquid spontaneously spreads on the solid surface. It is said to perfectly wet the solid surface. However, if the forces reach an intermediate balance, the liquid drop will establish a contact angle θ with the solid surface as shown in Figure A.1.

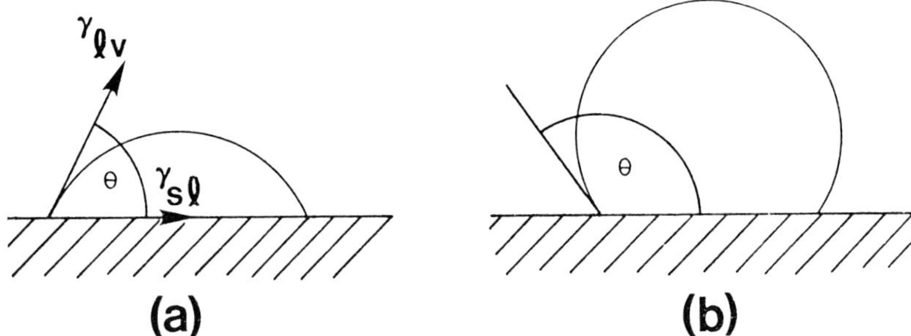

Figure A.1 Surface tension forces of an acute and obtuse contact angle for idealized liquid drops on a solid.

The balance of forces around the periphery of the drop where it meets the solid surface may establish an acute contact angle as in Figure A.1a, a 90° contact angle or an obtuse contact angle as in Figure A.1b. When the contact angle is less than 90° the liquid is said to wet the solid surface with the appropriate contact angle. When the contact angle is greater than 90°, nonwetting is said to occur. The balance of forces (or associated specific surface energies) may be expressed in this simplest case by Young's equation:

$$\cos\theta = (\gamma_{sv} - \gamma_{sl}) / \gamma_{lv} \tag{1}$$

where γ_{sv}, γ_{sl} and γ_{lv} are the interfacial tensions of the solid-vapor, solid-liquids, and liquid-vapor interfaces, respectively.

Equation (1) is derived either as a balance of surface tension forces as indicated in Figure A.1a or as a summation of the surface free energy changes which occur when unit area of the liquid is placed on unit area of the solid. Bikerman (7) objected to neglect of the vertical force component in Equation (1) but Lester (8) has shown that this component is important only for solid surfaces which are deformed appreciably by the liquid surface tension forces. Hard solids are apparently unaffected. Johnson (9) has derived Equation (1) on strictly thermodynamic bases.

The work of adhesion, W_{sl} between a solid and liquid is defined as:

$$W_{sl} = \gamma_s + \gamma_{lv} - \gamma_{sl} \tag{2}$$

where γ_s is the specific free surface energy of a clean solid in a vacuum. Note that γ_s differs from γ_{sv} because the surface energy of the solid has been decreased by adsorption of liquid vapor at a known partial pressure in the case of γ_{sv}. W_{sl} is defined as the work necessary to separate a unit area of solid-liquid interface into unit area each of liquid-vapor interface and clean solid surface. The adsorption of the vapor interface and clean solid surface. The adsorption of the vapor on the solid surface gives rise to a film pressure π_e defined as:

$$\pi_e = \gamma_s - \gamma_{sv} \tag{3}$$

Equations (2) and (3) may be combined to give an expression useful in the field of adhesion:

$$W_{sl} = \gamma_{lv} (1 + \cos \theta) + \pi_e \tag{4}$$

Often π_e is small and may be neglected when low vapor pressure liquids are involved. Equation (4) probably becomes invalid when $\theta = 0$.

Penetration of liquids into porous solid surfaces

When a liquid contacts a porous solid, the liquid surface bridging the periphery of pores generally becomes curved due to the differential tensions caused by the intermolecular forces of the liquid and solid. A pressure difference, ΔP, occurs across the curved liquid surface as shown in Figure A.2.

The pressure difference may be calculated from the LaPlace equation:

$$P_1 - P_2 = \Delta P = \gamma_l (1/r_{41} + 1/r_2) \tag{5}$$

where P_1 and P_2 are the pressures on the concave and convex sides of the meniscus, respectively, r_1 and r_2 are the principle radii of curvature of the liquid surface, and γ_l is the liquid surface tension in dynes/cm. The curvature and surface tension cause the liquid to spontaneously penetrate the pores, Figure A.2-a, or to resist penetration, Figure A.2-b, depending upon the algebraic sign of the curvature. Pores of irregular shape whose surfaces differ in composition from one area to another may cause the liquid surface to be *anticlastic* (saddle-shaped). In this case r_1 and r_2 are on opposite sides of the liquid surface and are opposite in algebraic sign. *Synclastic* liquid surfaces are produced when the pores are uniform in shape and composition and bokh r_1 and r_2 occur on the same side of the curved liquid surface. When the pores are cylindrical in cross-section the liquid surface becomes an element of a sphere and $r_1 = r_2 = r_o$. Equation (5) then becomes:

$$\Delta P = 2 \gamma_l / r_o \tag{6}$$

where r_o is the radius of the spherical liquid surface. If the liquid wets the solid with a finite contact angle θ, a cosine factor of the surface tension must

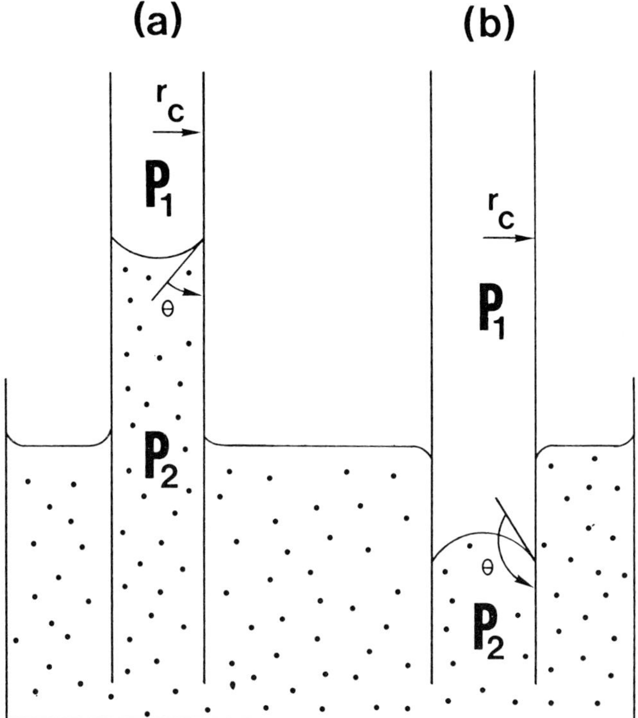

Figure A.2 Curved liquid surfaces in cylindrical pores of different surfaces.

be introduced and Equation (6) becomes:

$$\Delta P = \frac{2 \, \gamma_1 \, \text{Cos} \, \theta}{r_c} \tag{7}$$

where r_c is the radius of the cylindrical pore.

Equation (7) contains the fundamentals of sizing. It implies that as long as ΔP is positive the liquid should spontaneously penetrate the pore. However, if the contact angle between liquid and solid becomes 90°C, ΔP is zero and the liquid will not penetrate by capillarity. A hydrostatic head may be applied to the liquid to induce penetration if desired but no spontaneous penetration should occur. If θ becomes obtuse, then ΔP becomes negative and ΔP actually resists penetration of liquid into the pore. This would be expected to produce a high degree of sizing. In this case, if penetration were desired (such as in size pressing) it would be necessary to aplly an outside hydrostatic pressure on the liquid equal to or greater than ΔP in order for liquid to enter the pores. It is evident from Equation (7) that changes in either γ_1 or r_c will influence the magnitude to ΔP but only the Cos θ can change the algebraic sign. The sign of Cos θ is determined by the relative values of γ_s and γ_{sl} as shown in Young's equation or by suitable values of roughness and composition as indicated in Equations (23) and (24).

Equations (5-7) simply indicate whether the liquid will spontaneously penetrate or resist penetration into the porous structure. In many paper and board products it is unnecessary or even undesirable to make the product actually repellant to liquid. For example, recorder chart papers must have an ink wettable surface or the ink will not transfer from the pen nib to the paper. However, when the ink is transferred, the paper must resist feathering or line broadening. The rate of penetration of liquid is important for a majority of paper products. Combining Equation (7) with Poiseuille's equation for the volume rate of flow through a uniform bore capillary tube

$$V = \Delta P r_c{}^4 t / 8\eta l \tag{8}$$

where V is the volume of liquid of viscosity η flowing in time, t, under a pressure difference, ΔP, through a cylindrical tube of length, l, and radius, r, will give the Lucas-Washburn equation:

$$l^2 = \frac{\gamma_l\, r_c\, t\, \text{Cos}\,\theta}{2\eta} \pm \frac{r_c{}^2 dg}{8\eta} \tag{9}$$

In Equation (9), d and g are the density of the liquid and the acceleration due to gravity. When gravitational effects may be neglected, Equation (9) may be differentiated to give:

$$\frac{dl}{dt} = (\gamma_l r_c\, \text{Cos}\,\theta) / 4\eta l \tag{10}$$

where dl/dt is the rate of penetration due to capillarity. The limitations of this equation have been discussed by a number of investigators (54-58). Recent studies of Equation (10) have been made for organic liquids in glass capillaries by Good (58) and Good and Lin (59), for glass fiber filter papers by Everett and coworkers (60) and for papers by Schubert (61). A thorough review of the transudation of water into paper has been published by Hoyland (62). He has also considered the effects of swelling during penetration of aqueous liquids into paper (63).

Mechanisms of liquid pentration of paper

It is commonly assumed that liquids penetrate paper and board by capillarity. Probably this mechanism occurs to a degree in all methods of testing of sizing. However, the end points as determined by some size appear to depend upon different principles. Nissan (64) suggested in 1949 that four possible mechanisms exist. These are:

1. Liquid movement through the pores by capillary action.
2. Vapor phase movement through the porous structure.
3. Liquid movement through pores by surface diffusion.
4. Liquid movement through the solid fraction of the fibers.

It appears that different size tests may favor one mechanism over another depending upon the limiting mechanism and method of determining the end point.

The data of Verhoef, Hart and Gallay (65) show that dilute electrolytes migrate faster through sized paper *via* the solid fraction of the sheet than by capillary action. Reaville and Hine (66) have also considered noncapillary migration in paper. It has long been known that the TAPPI Size Test is unsuitable for measuring sizing in a hard-sized sheet because the vapor absorption by the indicator powder gives an end point before liquid water has moved through the sheet. The importance of vapor transmission in hot and cold drinking cup stock and in molded pulp floral containers and buckets in also well-known.

In summary, there are three requirements for good internal sizing:

1. The chemical composition of the sizing material must be hydrophobic (low surface energy).
2. The sizing material must be uniformly and completely retained on the fiber surfaces.
3. During drying the retained size must attain a stable low surface energy in the presence of the liquid to which the paper or board will be subjected during its useful life.

Contact angles

It is evident from the discussion above and from Equations (1), (7) and (10) that the major factor which governs wettability and sizing of paper and board is the *effective* value of the contact angle θ. The word effective is used here because, as later discussion will show, several contact angles may be defined depending upon the roughness and heterogeneity of the solid surface of interest. Under conditions of use, if the effective value of θ in Equation (10) is acute, the paper surfaces in contact with the size test liquid are wettable and a finite rate of penetration occurs. However, if θ is equation to $90°$ ($\mathrm{Cos}\ \theta = 0$) or greater than $90°$ (obtuse angle and $\mathrm{Cos}\ \theta$ is negative) then the surface is not wettable and penetration by capillarity should not occur. (It should be pointed out that some penetration may occur if sufficient *hydrostatic* pressure were to be applied, such as, in the nip of a size press, a blade coater, or during testing of sizing in, for example, a Penescope). The effective contact angle may also provide useful information for analysis of sizing problems, such as, fugitive sizing, and self sizing (2), and certain adhesion problems (3,32).

Measurement of contact angles

Methods of measuring contact angles between liquids and solids have been described by Adamson (18). Several of the more convenient methods involve the measurement of the angle which a tangent, drawn to the point of contact between the liquid and solid surface, makes with the solid surface. This is illustrated in Figure (A.1). The angle is often measured in a Contact Angle Goniometer. The use of liquid solutions for contact angle measurements is discouraged by both Zisman (39) and Good (25) because of adsorption uncertainties of components at the several interfaces. Pure liquids should be used. Paper surfaces often present special problems some of which are discussed in TAPPI Standard T458 m 59, and by Ekwall and coworkers (19, 20), Back and Lundin (21) and Kastner (22). In general, contact angle measurements on paper and paperboard are complicated by: surface roughness, surface heterogeneity of composition, moisture adsorption and adsorption, and changes in surface composition with time. Some of these effects are discussed in the section on contact angle hysteresis.

Equation (1) shows that the contact angle for a smooth planar surface is determined by the relative values of the three surface and interfacial tensions. It is clear that the values of γ_{sv} and γ_{sl} determine whether θ is acute or obtuse and therefore determine the wettability. If γ_{sv} is greater than γ_{sl}, θ is acute and wettability and poor sizing is produced. If γ_{sv} is less than γ_{sl}, θ is obtuse and good sizing is obtained. Thus, the sizing composition on the fiber surfaces of paper or board must possess a lower solid-vapor surface tension (or specific surface free energy) than the solid- liquid interfacial tension (or specific solid-liquid surface free energy). Because the size test liquids of interest are generally aqueous in nature and quite polar in composition, γ_{lv} is relatively large. The sizing composition should be nonpolar and its γ_s much smaller than γ_{lv} in order for γ_s to be smaller than γ_{sl}. Internal sizing toward oils and greases depends on the same principles; the only sizing compounds with γ_s less than the γ_{lv} of oils and greases are the fluorocarbons and certain silicones.

Experimental determination of γ_s or changes in γ_s of paper surfaces can be a useful indication of the chemical composition of the solid surfaces.

Determination of solid surface energy

General methods for determination of γ_s are not yet available, but there are two methods which can be used to relate chemical composition of solid surfaces to wettability. The first is an application of Zisman's concept of critical surface tension (10, 11) and the second method employs the Good-Girifalco-Fowkes' relationship between chemical composition and wettability.

The Zisman method for solid surface characterization

Zisman and coworkers (11) have shown that solid surface composition is relatively empiracally to liquid surface tension and their contact angle by an equation of the form:

$$\text{Cos } \theta = 1 + b \, (\gamma_c - \gamma_{lv}) \tag{11}$$

where b is a constant having a value in the order of 0.03 and γ_c is the critical surface tension of the solid. The critical surface tension of a smooth, solid surface can be determined by measuring the contact angles of a series of suitable liquids on the solid and plotting the Cos θ values *versus* the surface tensions of the liquids. The γ_c value is the surface tension at the intercept of the straight line plot with Cos $\theta = 1$. Liquids which have surface tension values equal to or less than γ_c have a zero contact angle and, therefore, wet and spread on those solid surfaces. Critical surface tensions obtained with the same series of liquids are related to surface chemical composition, degree of packing of the molecules and to molecular reorientation. Critical surface tension values vary from from 6 dynes/cm for $-CF_3$, 18 for Teflon, 20-22 for $-CH_3$ 31 for waxes, and polyethylene, 39 for starch, to 45 for cellulose. More extensive data are contained in (11). Zisman demonstrated that the attractive intermolecular forces at interfaces are controlled by only the surface atoms and molecules. Thus, a knowledge of γ_c and the surface tension of a liquid, γ_{lv}, permit prediction of wettability. The technique has been used for surface energy characterization of cellulose and wood components (12-17) and for predicting adhesion behavior in paper systems (3, 4, 5).

The Good-Girifalco-Fowkes' Method for solid surface Characterization

Girifalco and Good (23, 24, 25) proposed in 1957-1960 that the interfacial energy between two immiscible phases could be determined from the respective surface energies and the nature of the molecular attraction at the interface. The equation for a solid-liquid system is written:

$$\gamma_{sl} = \gamma_s + \gamma_{lv} - 2\,\phi\sqrt{\gamma_s\,\gamma_{lv}} \tag{12}$$

The terms are the same as previously defined except for ϕ which is the interaction parameter for the molecular attraction at the interface between the solid and liquid. Good (25) has shown that ϕ can be quantitatively expressed in terms of the molecular polarizabilities, dipole moments and ionization energies of the liquid and solid. For nonpolar liquids ϕ is approximately unity, but values between 0.5 and 1.0 are often encountered.

Combining Equations (1), (3), and (12) to eliminate ϕ gives:

$$\text{Cos } \theta = -1 + 2\,\phi\,(\gamma_s/\gamma_{lv})^{\frac{1}{2}} - \pi e/\gamma_{lv} \tag{13}$$

For liquids of similar ϕ and low enough vapor pressure so that ϕ is negligible, a plot of Cos θ *versus* $1/(\gamma_{lv})^{\frac{1}{2}}$ is linear. γ_s can then be determined from the slope of the line and the calculated value of ϕ. This yields a surface characterization of the solid. The limitations are discused by Good (25).

In 1964 Fowkes suggested (26) that the Girifalco-Good concept could be extended by assuming that the surface-free energies of solids and liquids are comprised of additive contributions from each type of intermolecular attraction present. For example, the surface-free energy of a polar solid may be comprised of several components:

$$\gamma_s = \gamma_s^L + \gamma_s^P + \gamma_s^H \ldots \tag{14}$$

where the superscripts L, P, and H denote the free surface energy components from the London (dispersion), polar, and hydrogen bonding attractive energies, respectively. The Debye induction effect is here assumed to be negligible. Fowkes also proposed that saturated hydrocarbons possesses only London-types of molecular interactions of the solids. One this basis Fowkes used a modified Girifalco-Good equation of the following form:

$$\mathrm{Cos}\ \theta = -1 + 2\ (\gamma_s^L / \gamma^*_{lv})^{1/2} - \pi e / \gamma^*_{lv} \tag{15}$$

where γ^*_{lv} is the surface energy of a completely nonpolar liquid. When πe = O plot of Cos θ versus $1/(\gamma^*_{lu})^{1/2}$ gives a straight line as shown in Figure A.3.

The slope of the line yields γ_s^L. Also the London components of polar liquids were obtained from measurements of the interfacial tension of the liquid against a liquid hydrocarbon using the following equation:

$$\gamma_{12} = \gamma_1 + \gamma_2 - 2(\gamma_1^L \gamma_2^L)^{1/2} \tag{16}$$

where γ_{12} is the interfacial tension between liquids 1 and 2 and γ_1 and γ_2 are the surface tensions of liquids 1 and 2, and γ_1^L and γ_2^L are the London components of liquids 1 and 2, respectively. The difference, $\gamma_{lv} - \gamma_1^L$ is attributed to all polar contributions γ_1^P. Tables of γ_s^L and γ_1^L values for low energy solids and common liquids are published (27, 28). There is a similarity between the γ_c values of Zisman and the γ_s^L values for the same solid.

The Owens and Wendt Method for solid surface characterization

Fowkes' reasoning has been extended by Owens and Wendt (30) and Kaeble (28, 29) to a treatment which gives the London and polar components of solid surface tension. This equation has the form:

$$\mathrm{Cos}\ \theta = -1 + 2(\gamma_1^L \gamma_s^L)^{1/2} / \gamma_{lv} + 2(\gamma_1^P \gamma_s^P)^{1/2} \gamma_{lv} \tag{17}$$

where $\gamma_1^L, \gamma_s^L, \gamma_1^P, \gamma_s^P$ are the London and polar components of the liquid and solid surface tensions, respectively. It is assumed that πe is negligible

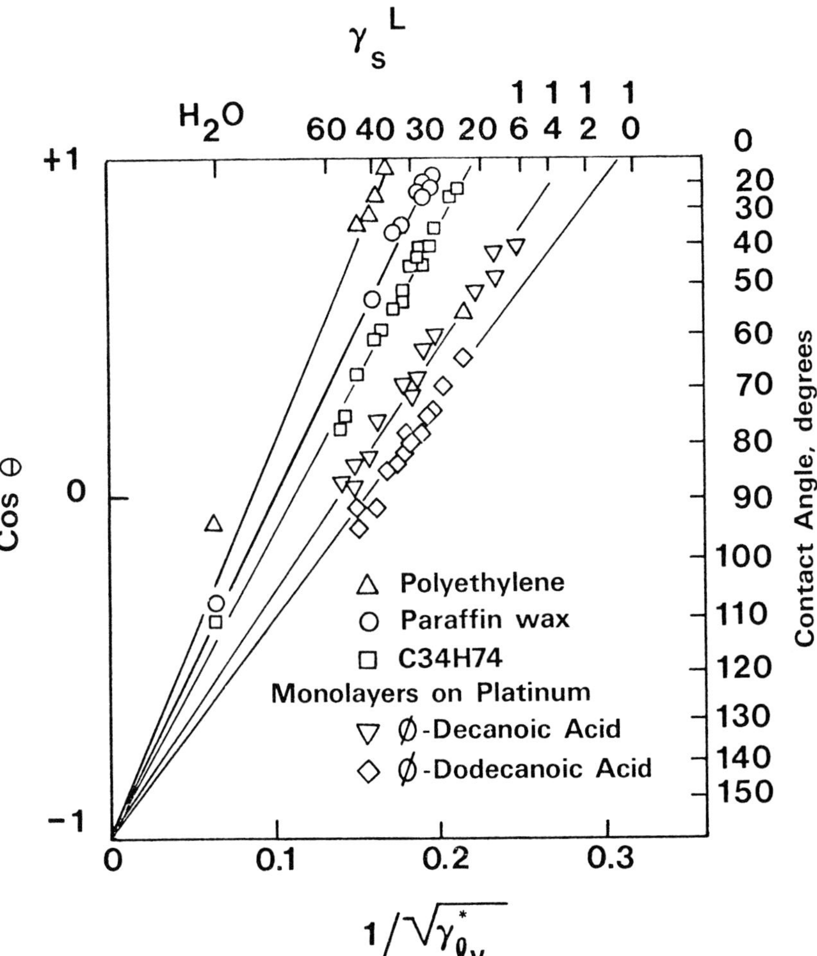

Figure A.3 Contact angles of a number of liquids on five low energy surfaces (27).

and that the surface tensions of the solid and liquids are comprised only of London and polar forces. It is further assumed that:

$$\gamma_s^L + \gamma_s^P = 1.0 \text{ and} \tag{18}$$

$$\gamma_l^L + \gamma_l^P = 1.0 \tag{19}$$

Equation (17) permits calculation of γ_s^L and γ_s^P from contact angles of two suitable liquids for which γ_l, γ_l^L γ_l, and γ_l^P are known. This gives two equations and two unknowns which may be solved by the method of simultaneous equations. The surfaces of cellulose, hemicellulose, lignin, and wood have been characterized by this method (15-17, 31, 32) as has a number of other low energy surfaces (33-38). This was done by Brown and Swanson (32) for cellulose films before and after treatment in a corona discharge. Data are shown in Table A.1.

Table A.1 Surface energy components of cellulose before and after corona discharge treatment.

| | Valve, ergs/cm^2 | |
Surface Component	Before Corona	After Corona*
London, γ	35.6	33.8
Polar, γ_s^P	23.2	32.8
Total, γ_s	58.9	66.6
Fractional polarity, γ_s^P / γ_s	0.40	0.49

Some comments on the methods

Controversies exist about the accuracy and validity of these methods for surface characterization. Wu (40) suggested that a reciprocal mean instead of the geometrical mean of Girifalco and Good (23, 24) better characterizes the London and polar interfacial interactions of polar polymers. The lumping together of the various polar types of contributions (e.g., dipole-dipole, dipole-induced dipole, hydrogen bonding, etc.) into one polar contribution may not be permissible. Andrade and coworkers (100) have adapted these techniques to measure polymer-water interfaces.

Also real solid surfaces are seldom smooth, planar, and isotropic as required by theory. Some of these complications are considered in the section on contact angle hysteresis.

High energy surfaces with γ_s greater than 100 ergs/cm^2 (e.g., metals, metal oxides, silicates, and glasses) often show autophobicity in contact with amphipathic (polar- nonpolar) liquids. Zisman and coworkers (41, 42) suggest that the liquid initially spreads on the solid surface. The solid then causes orientation of the first layer of molecules to a condition of sufficient low energy that the liquid cannot wet its own monolayer. The liquid then recedes and develops a finite contact angle.

High energy solids are more susceptible to contamination by adsorption from the atmosphere and other sources. The effects of moisture adsorption on contact angles of waterdrops on cellulose has been studied by Borgin (43). It is interesting that the contact angle of waterdrops on cellulose is a function of the relative humidity of the air, the water content of the cellulose, temperature, and time. For example, the contact angle decreased from an initial value of 30° to about 11° in 10 minutes. A contact angle of 25° was not lowered by increased moisture contents from 0 to 12% (corresponding to about 50% relative humidity, RH) but decreased steadily over the range of 25° to 12.5° at higher moisutre content from 12 to 20% (95% RH).

Dynamic wettability

It is apparent from the foregoing discussion that considerable attention has been given to equilibrium wettability of solid surfaces by liquids. However, it is equally apparent that the application of liquids to paper under conditions of industrial practice rarely allows time enough for thermodynamic equilibrium. Size pressing, surface sizing, coating, impregnation, and printing all involve treatment of the paper or board surface with liquids and colloidal dispersions at high speeds. The interactions of liquid and solid often take place in the millisecond range. In view of importance of dynamic wettability it is surprising that so little attention has been given to this subject.

The foregoing discussion of contact angles indicates that metastable states of the three phase line (TPL) occur depending on the vibrational energy of the liquid drop, molecular overturning, types of roughness, heterogeneity of surface composition, and penetration into pores. Consequently, the dynamic contact angle, defined as an instantaneous or steady state angle obtained during motion of TPL, is fundamentally different from the equilibrium contact angle. However, treatment of dynamic contact angles in terms of fluid dynamics has led to useful information.

Dynamic contact angles

There are two general cases of interest for study of dynamic wettability:

1. Spontaneous spreading of liquid film on the solid surface under the driving force of the imbalance expressed by Young's Equation (1), $\gamma_{sv} - \gamma_{sl} \geq \gamma_{lv}$.
2. The second case involves impressed motion of the liquid-vapor interface relative to the solid.

The first case is important in lubrication, hot melts, printing inks, foliar applications of herbicides and pesticides, and application of molten waxes, hot melts, printing inks, and organic films to paper and board. The influence of dynamic contact angles on spontaneous spreading of molten polymers (67-70) and oils (71) on solids has been studies by several groups of investigators. Capillary penetration occurs in voids when the liquid forms a contact angle less than 90°. Dynamic penetration into voids of solids has received much attention in textiles and underground oil recovery. Thorough coverage of this field has been published (72, 73).

The second case of dynamic wettability, that of imposed motion involves the increase of the advancing dynamic contact angle when the forced velocity is greater than the movement of the TPL. The early studies of Ablett (74) and Taggert and coworkers (75) showed a dependence of the dynamic contact angle with velocity of the moving liquid front. Advancing angles increased beyond the equilibrium contact angle and the receding angles decreased with increasing speed. Later studies by several groups of workers (76-81) confirmed a dependence of dynamic contact angle on velocity but there appears to be a lack of quantitative agreement in the results of one group of investigators to another. The data of Elliott and Riddiford (78) and Schwartz and Tejeda (81) suggest that factors which control the velocity-dynamic contact angle behavior are different at different speeds. On the other hand, Johnson and coworkers (82) show little or no dependence of dynamic contact angle on velocity of the three phase boundary.

Hansen and Miotto (83) have suggested that displacement of the TPL at higher speeds involves molecular disorientation in the immediate vicinity of the interface. The most energetically favorable orientation of the liquid and perhaps solid molecules at the TPL requires a finite time to achieve. If the displacement velocity is greater than the natural velocity of molecular movement, molecular disorientation develops and the interfacial free energies exceed their equilibrium values. However, at displacement velocities less than the natural velocity of molecular movement, the molecules at the interface should be able to orient themselves, and the dynamic contact angle should equal the equilibrium value. Other investigators (78, 80) have explained their results in terms of Hansen and Miotto's hypothesis. Lowe and Riddiford (84) propose that dynamic contact angle measurements can be used to determine the relative importance of London and polar molecular forces at the solid-liquid interface. They conclude that both polar and London forces are important at low velocities and that London forces dominate at higher velocities.

McIntyre (85) used an oscillating drop technique. He concluded that the energy differences related to the advancing and receding solid-liquid interface resulted from molecular relaxation effects in the three phase boundary region. Calculation of relaxation times for water molecules on paraffin gave realistic values of 10^{-6} to 10^{-5} sec.

Johnson and coworkers (82) found that velocities of up to 250 mm/min did not appreciably affect dynamic contact angles on homogeneous stable solid surfaces. However, heterogeneous solids showed a significant dependence on velocity.

Most of the work done on dynamic contact angles has been done at modest velocities up to a few hundred millimeters per minute on smooth, homogeneous surfaces. In spite of simplified systems there is currently a lack of agreement about the effect of velocity on dynamic contact angles and a proven theory is lacking. Significant advancement will probably come from a thorough molecular analysis of the three phase zone itself (83, 86) and a fluid dynamics approach (87, 88).

Dynamic wetting on paper and board

Several investigators have examined the rate of wetting during application of liquids to paper and board surfaces. Knight (89) applied colored liquids to paper surfaces from a moving slot followed immediately by a vacuum slot which removed liquid that did not wet the surface. He found that papers differ much in surface wettability. All papers will not accept aqueous liquids at sufficiently short time intervals or high enough speeds. This instrument is useful for examination of coating and bonding problems caused by surface heterogeneities. Recent scanning electron micrographs and movie films have illustrated the irregular and erratic wetting behavior of fibrous structures on a fine scale (90).

Bristow (91) studies dynamic wetting and penetration of paper by mounting a paper strip on a rotating wheel fixed with an applicator box resting on the paper. The box was filled with a known amount of liquid which was applied to the paper over a time determined by peripheral speed and the width of the box opening. This permitted determination of a roughness index and an absorption coefficient for paper from the following relationship:

$$V = K_r + K_a t^{1/2} \tag{20}$$

where V is the volume of liquid applied in time, t, and K_r is the roughness index and K_a is the absorption coefficient. K_r may be considered to be the mean depth of fluid penetration before capillary forces become effective. The effects of calendering pressure and surface finish were studied. It was noted that oils wetted the surface immediately and gave realistic K_r values whereas water required a short time before penetration into the sheet occurred. Bristow suggested that the wetting time for water was a two-stage process involving first, the time for the water to fill the surface roughness and secondly, the wetting time of the fiber surfaces.

Workers at the Institute of Paper Chemistry (92) constructed a blade applicator which permitted studies similar to Bristow's with various liquids and pigment coating mixtures. The total volume, V, of liquid applied is then the sum:

$$V + V_R + V_P + V_M \tag{21}$$

where V_R, V_P, and V_M are the voluems due to roughness, capillary penetration, and blade metering, respectively. Inserting the functions for V_P and V_M gives

$$V = VR + 2 \frac{K' \gamma_1 t \cos \theta^{1/2}}{\eta} + K \left(\frac{nv}{F}\right)^a \tag{22}$$

where K' is a permeation constant containing the void volume, the equivalent capillary size and the Darcy permeability constant, k and a are constants and F is the force per unit length of the metering blade. Operation of the coating process under blade pressure conditions where V_M is negligible and Cos θ is constant should give a linear plot of V *versus* $(\gamma_1 t/\eta)^{1/2}$ with an intercept at V_R and a slope of $2(K' \text{Cos } \theta)^{1/2}$. The relationship was obeyed with oils but not very well with aqueous solutions possibly due to high dynamic contact angles or to a time lag for initial wetting.

Other workers have studied liquid migration of coating fluids by optical methods (93, 94) and gravimetric techniques (95). Wink and Van den Akker (96) have constructed an apparatus for metering a fixed volume of liquid onto paper or board for determination of liquid permeability and surface roughness. The theory and calculational procedures of this method have been discussed and improved by Hung, Nelson, and Van Eperen (97). These techniques have been useful for characterizing differences between papers and boards with regard to surface porosity and roughness but have contributed little to a fundamental understanding of high speed wetting. However, perhaps they furnish a starting point for further research.

Contact angle hysteresis

In theory Young's equation predicts a single well-defined contact angle between a pure liquid and a smooth, plane, nondeformable solid surface of uniform and contant chemical composition. A problem arises because real solid surfaces seldom satisfy these requirements. This leads to *contact angle hysteresis,* a condition which permits a number of stable equilibrium contact angles to occur for a given solid-liquid system. An example is shown in Figure A.4 where it is observed that a larger contact angle θ_a is obtained when the volume of liquid is increased before measurement as compared with the lower contact angle θ_r obtained when the volume has been decreased. θ_a is called the advancing contact angle and θ_r the receding contact angle. Other examples and detailed discussion are given by Johnson and Dettre (44), Blake and Haynes (45), and Schwartz (46). A symposium on the subject has been edited by Padday (47).

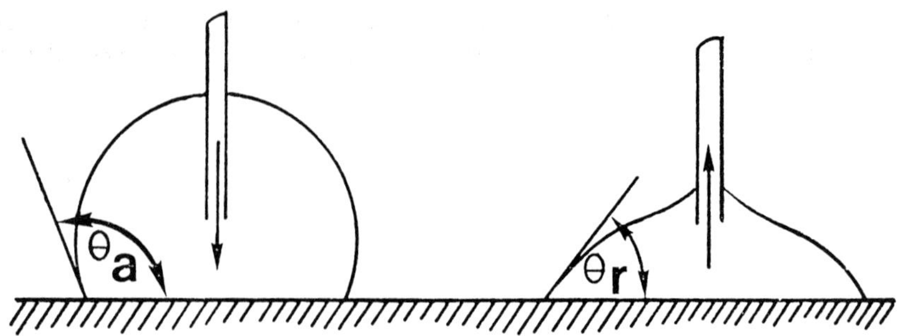

Figure A.4 Technique for obtaining advancing and receding contact angles (44).

The causes of contact angle hysteresis are generally attributed to changes in the solid which cause it to deviate from the theoretical fixed single composition, and planar, nondeformable surface. Surface roughness, heterogeneity of composition, molecular overturning, and porosity all appear to cause contact hysteresis.

Effects of surface roughness on contact angles

Wenzel (48, 49) has described the effects of surface roughness on wetting in terms of two contact angles, the observed or apparent angle θ_{obs}, the real or intrinsic angle, θ_o, and a roughness factor, r. The relationship between these quantities may be stated as:

$$\text{Cos } \theta_{obs} = r \text{ Cos } \theta_o \qquad (23)$$

where the roughness factor, r, is defined as the ratio of the real surface area divided by the geometric surface area. Since all real surfaces have areas greater than the geometric surface area $r > 1.0$. Equation (23) implies that when the intrinsic contact angle, θ_o, (on a smooth planar element of surface) is acute, roughness of the surface decreases θ_{obs}, and improves wettability of the surface. Also when θ_o is $>90°$, surface roughness increases $>_{obs}$ to an obtuse value and decreases wettability. The general effect is shown in Figure A.5 for increasing surface roughness ratios. It is observed that the effects of roughness are greater at very low and very high intrinsic contact angles. This implies that the surface must be much smoother for accurate measurements when θ_o is small than when θ is larger. Although Wenzel's analysis is illuminating, Johnson and Dettre (44) point out that it does not take into account the metastable states of the surface produced by roughness.

Cassie and Baxter (50, 51) derived a more general form of Wenzel's equation and applied it to woven and other porous structures. Johnson and Dettre (44, 52, 53) have presented an exhaustive study of the relationships between roughness and contact angle. This work has implications for sizing of paper and board where the surface may be rough, directional and porous, and the surface compsition may be variable.

The Cassie-Baxter equation for a composite surface of regions of different wettability is:

$$\text{Cos } \theta_{obs} = f_1 \text{ Cos } \theta_1 + f_2 \text{ Cos } \theta_2 \ldots \qquad (24)$$

where f_1 is the fraction of surface having a contact angle θ_1 and f_2 is the fraction having contact angle θ_2 and $f_1 + f_2 = 1$. At large roughness the liquid may not penetrate to the valleys between peaks but may bridge over them. Under these conditions Equation (24) reduces to:

$$\text{Cos } \theta_{obs} = f_1 \text{ Cos } \theta_1 - f_2 \qquad (25)$$

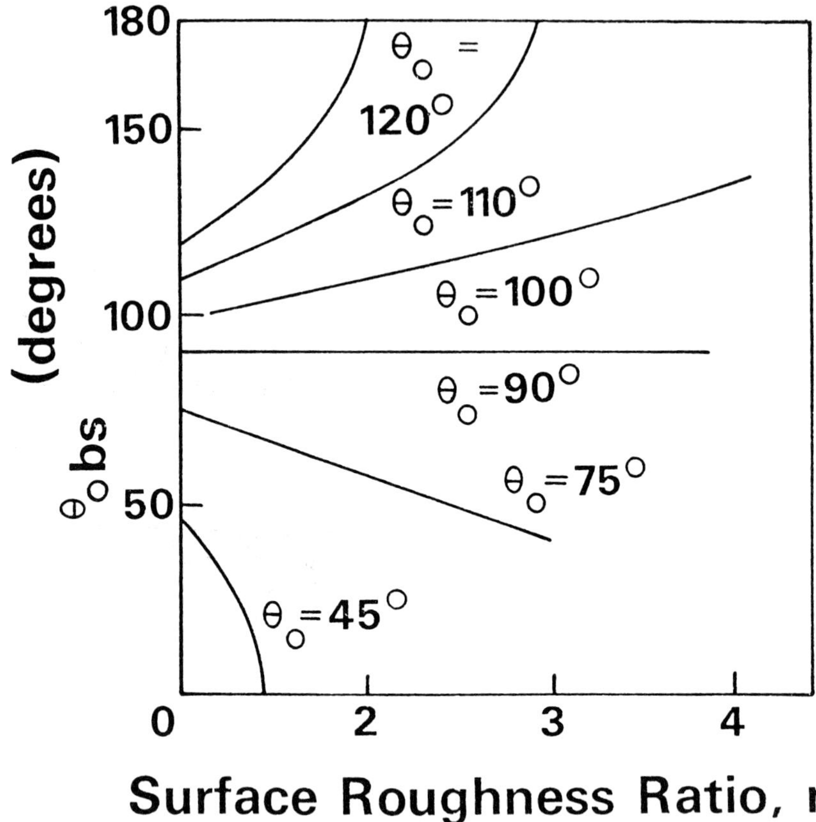

Figure A.5 Wenzel's angle as a function of surface roughness for various intrinsic angles.

As f_2, the liquid surface area over the voids, increases, the observed contact angle, θ_{obs}, becomes larger and the overall wettability of the surface becomes poorer. This implies that rougher surfaces are likely to be less wettable by adhesives or other liquids. Johnson and Dettre (44) have shown that both the advancing and receding contact angles approach the angle predicted by Cassie and Baxter's equation at larger roughness ratios. For surfaces comprised of patches of low and high surface energy material it is concluded that the advancing angle is a better measure of the wettability of the low energy (high contact angle) part of the surface and a receding angle is a better measure of the high energy part (52).

Effect of molecular overturning on contact angles

Langmiur, Hardy and Harkins had suggested much earlier that advancing and receding contact angles and variation of contact angles with time could be caused by molecular overturning. However, Yiannos (53) was probably the first to present data which offer support for such changes at the interface between solid and liquid. He developed both kinetic and thermodynamic expressions for the process of overturning of amphipathic monolayers on nonreactive surfaces. Contact angle data on long-chain fatty acid layers show that the rate of overturning in the presence of water is dependent upon temperature and chain lengths. The 18-carbon atom stearic acid monolayers required 55 hours at 13–15°C for the contact angle to decrease from an initial value of 105° to the equilibrium value of 70°. At higher temperatures of 23-28°C only 10–15 hours were required. Fatty acids of shorter chain length, myristic (14 C-atoms) and lauric (12 C-atoms), required minutes and seconds, respectively, to attain equilibrium contact angles. It was further shown that if the monolayer can be chemically reacted with the solid surface overturning is prevented and the contact angle remains at a constant obtuse value. This implies that a sizing agent for paper and board must not molecularly overturn in the presence of the size test liquid.

Literature cited

1. Patnode, W. I., U.S. Patent 2,306,222.
   Robbart, C. F., U.S. Patent 3,56,558.
   Robbart, C. F., Tappi 59:68(Oct. 1976).
2. Swanson, J. W. and Cordingly, S., Tappi 42:812 (Oct. 1959).
3. Swanson, J. W., and Becher, J. J., Tappi 49:198 (1966).
4. Glossman, N., Tappi 50:224(1967).
5. Fredholm, B. and Westfelt, L., Svensky Papperstidn. 82, no. 11:202(Aug. 1979).
6. Hirschfelder, J. O. ed., Intermolecular Forces. Interscience Publishers, New York, New York, 1967.
7. Bikerman, J. J. Contributions to the thermodynamics of surfaces. Massachusetts Institute of Technology, Cambridge, Mass.
   Bikerman, J. J., Proc. 2nd International Congress on Surface Activity, London, 1957,
8. Lester, G. R., J. Colloid Sci. 16:315(1961).
9. Johnson, R. E., J. Phys. Chem. 63:1655(1959).
10. Fox, H. W., and Zisman, W. A., J. Colloid Sci. 5: 514 (1950).
11. Zisman, W. A. Advan. Chem. Series No. 43, Chapter I, American Chemical Society, Washington, D.C., 1964.
12. Ray, B. R., Anderson, J. R., and Scholz, J. J., J. Phys. Chem. 62:72(1958).
13. Gray, V. R., For. Prod. J. 12:452-61(Sept. 1962).
14. Herczeg, A., For. Prod. J. 15:499-505(Nov. 1965).

15. Luner, P., and Sandell, M., J. Polmer Sci., Part C28, 115(1969).
16. Lee, S. B., and Luner, P., Tappi 55:116(1972).
17. Nguyen, T., and Johns, W. E., Wood Sci. Technol. 12:63(1978).
18. Adamson, A. W. Physical Chemistry of Surfaces. 3rd ed. John Wiley & Sons. New York, New York, 1976,
19. Ekwall, P., and Andresen, O., Paper and Timber (Finland) 40:173-176, 178-9, 181(Special Number April 1958).
20. Ekwall, P., Bruun, H. H., and Lahtonen, P., Paper and Timber (Finland 40:182-9(Special Number, April 1958).
21. Back, E., and Lundin, B., Svensk Papperstidn. 58:758(1955).
22. Kastner, H. U., Zellstoff Papier 27:116-121(May/June, 1978).
23. Girifalco, L. A., and Good, R. J., J. Phys. Chem. 61:904(1957).
24. Good, R. J., and Girifalco, L. A., J. Phys. Chem. 64:561(1960).
25. Good, R. J., J. Colloid Interface Sci. 59:398(1977).
26. Fowkes, F. M., J. Phys. Chem. 72:370(1968).
27. Fowkes, F. M. Chemistry and physics of interfaces. S. Ross, ed., p. 154, American Chemical Society, 1971.
28. Kaelble, D. H. Physical chemistry of adhesion. Chapter 5. Wiley-Interscience, New York, 1971.
29. Kaelble, D. H., and Uy, K. C., J. Adhesion 2:50(1970).
30. Owens, D. K., and Wendt, R. C., J. Applied Polymer Sci. 13:174(1969).
31. Ferris, J. L. The wettability of cellulose film as affected by vapor-phase adsorption of amphipathic molecules. Doctor's Dissertation. The Institute of Paper Chemistry, Appleton, Wis., 1974.
32. Brown, P. F., and Swanson, J. W., Tappi 54:201(1971).
33. Wu, S., J. Colloid Interface Sci. 31:153(1969).
34. Dann, J. R., J. Colloid Interface Sci. 32:321(1970).
35. Schwarcz, A., J. Polymer Sci. 12:1195(1974).
36. Kloubeck, J., J. Colloid Interface Sci. 46:185(1974).
37. El-shimi, A., and Goddard, E. D., J. Colloid Interface Sci. 48:242(1974).
38. Roitman, J., and Pittman, A. G., J. Polymer Sci. 12:1195(1974).
39. Zisman, W. A., J. Paint Technol. 44:42(1972).
40. Wu, S., J. Polymer Sci., Part C, 34:19(1971).
41. Hare, E. F., and Zisman, W. A., J. Phys. Chem. 59:335(1955).
42. Bernett, M. K., and Zisman, W. A., Adv. Chem. Series No. 43, p. 1, 332. American Chemical Society, Washington, D.C., 1964.
43. Bogin, K. The properties and nature of the surface cellulose. (In English). Part One: Theoretical and experimental. Norsk Skogind. 13:81-92(1959). Part Two: Cellulose in contact with water. Norsk Skogind. 13:429-44(1959).
Part Three: Cellulose in contact with organic liquids and aqueous solutions of reduced surface tension. Norsk Skogind. 14:485-495(1960). Park Four: The effect of changing or modifying the surface of cellulose. Norsk Skogind. 15:384-391(1961).
44. Johnson, R. E., Jr., and Dettre, R. H. *In* Matijevic's, Surface and Colloid Science. Vol. 2, p. 85, Wiley, New York, 1969.

45. Blake, T. D., and Haynes, J. M. *In* Danielli, Rosenberg, and Cadenhead's Progress in Surface Membrane Science. Vol. 6, p. 125, Academic Press, New York, 1973.
46. Schwartz. A. M., Adv. Colloid Interface Sci. 4:439(1975).
47. J. F., ed. Wetting, spreading and adhesion. Sections I and II, p. 3-124. Academic Press, New York, 1978.
48. Wenzel, R. N., Ind. Eng. Chem. 28:988(1936).
49. Wenzel, R. N., J. Phys. Colloid Chem. 53:1466(1949).
50. Cassie, A. B. D., and Baxter, S., Trans. Faraday Soc. 40:546(1944).
51. Baxter, S., and Cassie, A. B. D., J. Textile Inst. 36:T67(1945).
52. Dettre, R. H., and Johnson, R. E., Jr., J. Phys. Chem. 69:1507(1965).
53. Yiannos, P. N., J. Colloid Sci. 17:334-47(April 1962).
54. Washburn, E. W., Phys. Review 17:273, 374(1921).
55. Lucas, R., Kolloid Z. 23:15(1918).
56. Penetration of Papers by Liquids and Solutions. Instrumentation Studies XXXIV, The Institute of Paper Chemistry. Paper Trade J. 110-44-50(Jan. 25, 1940).
57. Van den Akker, J. A., and Wink, W. A., Tappi 52:2406(1969).
58. Good, R. J., J. Colloid Interface Sci. 42:473-477(March 1973).
59. Good, R. J., and Lin, J., J. Colloid Interface Sci. 54:52-58(Jan. 1976).
60. Everett, D. H., Haynes, J. M., and Miller, R. J. L., Trans. BPBIF Symp. Fiber-Water Interactions in Papermaking (Oxford) p. 519-536(Sept. 1977).
61. Schubert, H., Trans. BPBIF Symp. Fiber-Water Interactions in Papermaking (Oxford), p. 537-555(Sept. 1977).
62. Hoyland, R. W., Paper Technology and Ind.
    I. Some structural aspects, 17:213(Oct. 1976).
    II. The cellulose-water relationship, wetting and contact angles, 17:216-219(Oct. 1976).
    III. Some principles of flow and their application to paper, 17:291-293, 298-9(Dec. 1976).
    IV. Liquid penetration into paper, 17:304-6(Dec. 1976).
    V. The mechanism of pentration, and conclusions, 17:7-9(Jan. 1977).
63. Hoyland, R. W., Trans. BPBIF Symp. Fiber-Water Interactions in Papermaking (Oxford), p. 557-579(Sept. 1977).
64. Nissan, A. H., Proc. Tech. Sect. P.M.A., 30:96(1949).
65. Verhoef, J., Hart, J. A., and Gallay, W., Pulp Paper Mag. Can 64:T509(1963).
66. Reaville, E. T., Hine, W. R., Jr., Tappi 50:262(1967).
67. Schonhorn, H., Frisch, H. L., and Kivei, T. W., J. Appl. Phys. 37:4967(1966).
68. Van Oene, H., Chang, Y. F., and Newman, S., J. Adhesion 1:54(1969).
69. Yin, T. P., J. Phys. Chem. 73:2413(1969).
70. Cherry, B. W., and Holmes, C. M., J. Colloid Interface Sci. 29:174(1969).
71. Bascom, W. D., Cottington, R. L., and Singleterry, C. R., Advanc. Chem. Series No. 43, p. 355. American Chemical Society, Washington, D.C., 1964.

72. Scheidegger A. E. The physics of flow through porous media. Univ. of Toronto Press, Toronto, 1960.
73. Flow through porous media. American Chemical Society, Washington, D.C., 1970.
74. Ablett, R., Phil. Mag. 46:744(1923).
75. Taggert, A. F., Taylor, T. C., and Ince, C. R., Trans. AIME 87:285(1930).
76. Yarnold, G. D., and Mason, B. J. Proc. Phys. Soc., B62:125(1949).
77. Rose, W., and Heins, R. W., J. Colloid Sci. 17:39(1962).
78. Elliot, G. E. P., and Riddiford, A. C., J. Colloid Interface Sci. 23:389(1967).
79. Blake, T. D., and Haynes, J. M., J. Colloid Interface Sci. 30:421(1969).
80. Phillips, M. C., and Riddiford, A. C., J. Colloid Interface Sci. 41:77(1972).
81. Schwartz, A. M., and Tejeda, S. B., J. Colloid Interface Sci. 38:359(1972).
82. Johnson, R. E., Jr., Dettre, R. H., and Brandreth, D. A., J. Colloid Interface Sci. 62:205(1977).
83. Hansen, R. S., and Miotto, M., J. Am. Chem. Soc. 70:1765(1957).
84. Lowe, A., and Riddiford, A. C., Can. J. Chem. 48:865(1970); J. Chem. Soc. 1970D:397(1970).
85. McIntyre, D. E., A study of dynamic wettability on a hydrophobic surface. Doctor's Dissertion. The Institute of Paper Chemistry, Appleton, Wis., 1969.
86. Blake, T. D., and Haynes, J. M. *In* Danielli, Rosenberg, and Cadenhead's Progress in Surface and Membrane Sci., Vol. 6, p. 125, Academic Press, New York, 1973.
87. Scriven, L. E., Chem. Eng. Sci. 12:98(1960).
88. Huh, C., and Scriven, L. E., J. Colloid Interface Sci. 35:85(1971).
89. Knight, G. D., Tappi 33:59-66(Feb. 1950).
90. Oliver, J. F., and Mason, S. G., Trans. BPBIF Symp. of the Fundamental Properties Related to Paper Uses, Cambridge, P. 428(Sept. 1973, publ. 1976).
91. Bristow, J. A., Svensk Papperstidn. 70:623(1967).
92. The Instutute of Paper Chemistry, Project 2495, Report 5 (July 12, 1976).
93. Lyne, M. B., Transactions of BPBIF Symp. on Water-Fiber Interactions in Papermaking (Oxford), p. 604(1977).
94. Windle, W., Beazly, K. M., and Climpson, M., Tappi 53:2232(1970).
95. Rosen, A., and Hemstock, G. A., Tappi 50:7(Dec. 1967).
96. Wink, W. A., and Van den Akker, J. A., Tappi 40:528(1957).
97. Hung, J. Y., Nelson, R. W., and Van Eperen, R. H., Tappi 52:1732(Sept. 1969).
98. Desai, R. L., and Shields, J. A., J. Colloid Interface Sci. 31:585(1969).
99. McKelvey, R. D., and Leekley, R. M., J. Colloid Interface Sci. 56:377(Aug. 1976).
100. Andrade, J. D., *et al.*, J. Colloid Interface Sci. 72:488(Dec. 1979).

Index